D0359660

EARTH AT RISK

CLAUDE HENRY AND
LAURENCE TUBIANA

EARTH AT RISK

Natural Capital and the Quest for Sustainability

COLUMBIA UNIVERSITY PRESS
NEW YORK

Columbia University Press
Publishers Since 1893
New York Chichester, West Sussex
cup.columbia.edu

Library of Congress Cataloging-in-Publication Data
Names: Henry, Claude, author. | Tubiana, Laurence.
Title: Earth at risk : natural capital and the quest for sustainability / Claude Henry and Laurence Tubiana.
Description: New York : Columbia University Press, 2017. | Includes bibliographical references and index.
Identifiers: LCCN 2017026502 | ISBN 9780231162524 (alk. paper)
Subjects: LCSH: Sustainable development. | Natural resources—Management. | Global environmental change.
Classification: LCC HC79.E5 H463 2017 | DDC 338.9/27—dc23
LC record available at https://lccn.loc.gov/2017026502

Columbia University Press books are printed on permanent and durable acid-free paper.
Printed in the United States of America

Cover design: Noah Arlow

This is only the first sip, the first foretaste of the bitter cup which will be proffered to us year by year, unless by a supreme recovery of moral health and martial vigour, we arise again.

Winston Churchill, *On the Munich Agreement*,
House of Commons, October 5, 1938

CONTENTS

ACKNOWLEDGMENTS

BOTH KEN ARROW and Tony Atkinson are present in the book, and to both we owe illuminating suggestions as well as friendly encouragement; neither is still with us, albeit they remain deep in our hearts.

This book would not have been possible without the interactions with our students at Columbia University, the Institut d'Etudes Politiques de Paris, and Ecole Polytechnique Paris. They had previously graduated in all kinds of subjects, and they were all new to the approach taken here. That required us to make constant effort to clarify both our conceptual clarity and our pedagogy, as well as to achieve an appropriate balance among conceptual developments, illustrations, and applications. These efforts should make the book accessible and attractive to larger audiences of readers interested in the issues discussed.

We are also much indebted to Michel Balinski, Pierre Barthélemy, Jean-Pascal Benassy, Michel Berry, Isabelle Biagiotti, Dominique Bureau, Damien Conare, Michel Colombier, Pierre Ducret, Anne-Laure Faure, Emmanuel Guerin, Alain Grandjean, Geoffrey Heal, Jean Jouzel, David Kennett, Jean-Christian Lambelet, Yann Laurens, Valérie Masson, François Moisan, Pippo Ranci, Vincent Renard, Nicholas Stern, Thomas Sterner, Sébastien Treyer, and Tancrède Voituriez.

Support is gratefully acknowledged from the Institut du Développement Durable et des Relations Internationales (Paris), with Lisa Dacosta and Laetitia Dupraz; from the Chaire du Développement Durable at Ecole Polytechnique (Paris), with Lyza Racon; and from the Alliance Programme (New York and Paris), under Alessia Lefébure.

At Columbia University Press, Bridget Flannery-McCoy and Marielle Poss were in charge of steering the manuscript toward its final form, and Ben Kolstad of Cenveo oversaw the production. We had not imagined how crucial their contributions would be. Bridget in particular suggested how to properly structure an initially disorganized text, and she helped prune unnecessary material. Ben not only clarified the formulation but also sought precision everywhere it was somewhat lacking.

EARTH AT RISK

INTRODUCTION

WHY SHOULD we muster the best of our resources—both human and material—to implement a more sustainable form of development? The answer is that billions of our fellow humans live in extreme poverty, and this will not change without deeply changing the present mode of development. Moreover, the condition of our planet worsens at such a pace that all forms of life, including ours, will come under the most serious threats during the present century. Current generations are squandering, at an unbearable pace, the heritage of natural capital in their hands, and poor communities are suffering and will suffer most.

We start this introduction by considering the manifestations of this trend of unsustainable development. We then examine the array of tools—scientific and technological, institutional and legal, economic and managerial—that we might muster to reverse this trend. But will any of these tools actually be deployed on a proper scale, and the trend reversed? Right now, it is impossible to tell. It is nevertheless clear that a wealth of ongoing initiatives, which we'll consider later in this introductory chapter, points in the right direction, although those initiatives don't (yet?) have the collective strength and coherence to overcome the formidable obstacles in the way. In this respect, the 2015 Paris Agreement on climate change may, for all its imperfections, be seen as a pioneering institution:[1] a framework and a springboard for promoting, sanctioning, and coordinating all sorts of actions aimed at containing climate change.

NATURAL CAPITAL DEVASTATION

In an unusual statement on the eve of the 2012 Rio+20 Conference, the editors of *Nature* reminded their readers:

> The most unique feature of Earth is the existence of life, and the most extraordinary feature of life, is its diversity. Approximately 9 million types of plants, animals, protists, and fungi inhabit the earth. So, too, do 7 billion people. Two decades ago, at the first Earth Summit, the vast majority of the world's nations declared that human actions were dismantling the Earth's ecosystems, eliminating genes, species and biological traits at an alarming rate. This observation led to the question of how such loss of biological diversity will alter the functioning of ecosystems and their ability to provide society with the goods and services needed to prosper (Cardinale et al. 2012, 59).

Needless to say, the rate of dismantling is no less alarming today than it was two decades ago. Marine ecosystems are in such danger that it is no longer absurd to envisage biodiversity at sea reduced to the diversity of jellyfish varieties. Rain forests—under the pressures of palm oil production, soybean farming, beef ranching, wood smuggling, legal and illegal mining, pervasive greed, corruption, banditism, and even warfare—fare no better than the oceans. Nor are temperate forests immune to degradation; in northwestern America and across Siberia, fungi, beetles, and fires, whipped up by climate change, conspire in a mass extermination of trees. (We will consider biodiversity erosion further in chapter 1.)

Water and soil have in common an essential trait: both are subject to enormous waste. Irrigation accounts for two-thirds of the total quantity of water used by human beings. One-half of this, or fully one-third of the total, is lost. In many places, from Texas to Punjab, farmers pump as much water as they can from fast-declining aquifers, convinced that any drop they fail to extract will be pumped—expropriated, as they see it—by their neighbors.

Dominant approaches to soil management are just as reckless, as exemplified by the way nitrogen fertilizers are applied. In Europe and the United States, only about 20 to 30 percent of the amount of nitrogen applied to the fields is actually taken up by the plants. In China, it is even less. The surplus not only pollutes both air and water but also acidifies soils, which becomes toxic to crops. This is one of the main reasons, according to a report on November 4, 2014, by Xinhua, the official press agency of the People's Republic of China, why "more than 40% of China's arable land is suffering from degradation, seriously reducing the country's capacity to produce food for the world's biggest population." (More on water and soil waste in chapter 2.)

While scarcity of clean water and arable soil is a pending disaster, a resource for which greater scarcity would be welcome is fossil fuels. Scientists not only agree that carbon release is the main cause of global warming; most now say that to avoid major climate disturbances, the increase in the earth's mean temperature must be kept under 1.5°C–2°C (compared with preindustrial levels). This requires leaving at least two-thirds of all proven conventional fossil fuel reserves in the ground and not tapping shale, deep sea, or Arctic reserves. As the executive summary to the 2012 International Energy Agency's *World Energy Outlook* put it: "No more than one-third of proven reserves of fossil fuels can be consumed prior to 2050 if the world is to achieve the 2°C goal" (3). If this warning is ignored for too long, the damage shall not be confined to the environment; the reports from the Carbon Tracker Initiative (Leaton 2012) anticipate that when businesses finally have to reckon with the contradiction between short-term profitability anticipations and long-term global sustainability requirements, there shall be a "rupture" of historical proportions on financial markets, dwarfing 2008. (We go deeper into energy constraints and prospects in chapter 3.)

Despite the vital importance of biodiversity, water, soil, and decarbonized energy, climate will ultimately prove the critical component of natural capital. Climate change is often seen as a threat only for distant generations. But a look to the past can put things in perspective. Fifty-six million years ago, the continents were breaking apart, the earth crust was in flux, and such intense heat was released that carbon-rich sediments were baking and, in the process, emitting huge amounts of carbon dioxide (CO_2) and methane (CH_4). The earth's mean temperature increased by between 7°C and 8°C, and the temperatures of the ocean surface at polar latitudes reached between 20°C and 25°C, providing a comfortable environment for alligators. This so-called Paleocene-Eocene Thermal Maximum (PETM) event was short-lived (from a geological point of view), lasting "only" some thousands of years.

Now imagine you are watching the film *Rebel Without a Cause*. Jim Stark and Buzz Gunderson (memorably played by James Dean and Corey Allen), vying for the attention of Judy (Natalie Wood), are driving their cars toward the cliff in a game of chicken. That scene is gripping enough. But now think of accelerating the speed of the film twentyfold: it becomes mind-boggling. Nevertheless, that's what is happening with climate change today: the underlying rate of carbon emission is at least twenty times higher than that sustained during the huge PETM climate shift. (More on climate change in chapter 4.) If not reversed, the trends in natural capital devastation—climate change, in particular—will inevitably lead to an "unlivable planet" (to quote the *New York Times* editorial on December 25, 2015).

TOOLS FOR REVERSING THE TREND

Why, in these dire circumstances, does humankind hesitate, frozen like a guinea pig mesmerized by a boa? It's definitely not for want of ingenuity. There is an astounding range of knowledge and instruments—technological, economic, managerial, legal—that are available right now, or will soon be. To what extent can they help?

In some cases, tools provide the knowledge we need to make informed decisions. Consider biodiversity: being able to accurately measure the recovery rates of marine species in protected zones provides accurate guidelines and powerful arguments for barring broad abuses of ocean life resources. In the same spirit, properly assessing the range of carbon exchanges between trees and soils in tropical forests dramatically reinforces the case for preserving their integrity.

In other cases, we must make better use of what we have. Daniel Hillel's observations and innovations as a young scientist and engineer on a Negev kibbutz revolutionized the way water is brought to plants, securing spectacular water savings. About sixty years later, Hillel received the 2012 World Food Prize "for conceiving and implementing a radically new mode of bringing water to crops in arid and dry land regions, known as microirrigation" and for promoting it "in dozens of countries in Africa, Asia and Latin America" (Statement of Achievement of the Laureate), as well as other Middle Eastern countries. Soil is no less squandered than water. Is agriculture destroying the very resources on which it depends? This need not be a fatality, as advanced biological approaches prove efficient substitutes for noxious chemical ones.

This spirit of innovation is now trained on renewable sources of energy. Once fickle and intermittent, the reliability of renewables will be greatly enhanced as high-capacity devices for energy and electricity storage become available. New developments are thus overcoming the obstacles that once stood in the way of large-scale use of these technologies. The performance of solar panels was somewhat disappointing when first introduced; now we have thin films printed with photovoltaic cells that require far less material inputs and that yield 50 percent more energy than current panels. Are chemical reactions energy voracious? This is less and less the case, because quantum chemistry informs the selection of the appropriate catalysts for specific reactions, and catalysts are great energy savers. We are in a dynamic Schumpeterian mode of putting science to work. (In chapter 5, we'll explore the relevance of the scientific method to a transition toward a more sustainable mode of development. In chapters 6 and 7, we'll highlight some particularly valuable applications and extensions.)

Moving beyond natural sciences and technology: Are we hamstrung by markets that are too limited in scope to address the problems we face? If so, it is within our ingenuity and power to broaden the market mechanisms. Australian farmers, battered by years and years of drought, have finally conceded to governance reforms in the country's agricultural heartland—the vast Murray-Darling basin. They have accepted that rights to access water should be issued and traded on a specific market, for the sake of inducing more efficient allocations. Irrigation patterns have since changed, and thirsty crops have been substituted by less thirsty ones.

Tax reforms may also—even prominently—be used to redress pricing flaws. In Sweden, carbon is now taxed at about $150 per ton of CO_2 emitted. While this might seem mad to some, this price simply reflects the damages expected from CO_2 emissions. The tax is inducing increasingly sober resource use, without harming the combination of economic prosperity and social justice that is the country's signature.

Adequate combinations of technological, economic, and social approaches are not consigned to rich countries. In Bihar and Uttar Pradesh, two of the more populated and poorer states in India, a team of Indian innovators has combined technical ingenuity—in the spirit of Einstein's rule "As simple as possible, but no simpler"—with responsible economic approaches and responsive managerial skills to create Husk Power Systems. Their goal is to provide electricity to rural communities that never dreamed of being wired. The key is rice husk, an abundant, discarded, and renewable local resource that can be used as a fuel when gasified. The actors are local people who are trained to run a community plant or to supervise a cluster of plants. Such simple plants are burgeoning throughout the two states. This is a remarkable example of sustainable endeavor, attuned to essential needs and combining technical, managerial, and social innovations. (More on such innovations in chapter 8.) And it is by no means unique. There is indeed no shortage of investigations—in particular those by Elinor Ostrom and colleagues, which she brought together in her 2009 Nobel Prize lecture "Beyond Markets and States: Polycentric Governance of Complex Economic Systems." She showed that it's possible to overcome the tragedy of the commons and to achieve outcomes that fly in the face of narrow concepts of economic rationality, at least as the scale of communities whose members have been able to nurture sufficient trust.

A WEALTH OF INITIATIVES

But where one lacks trust, shared goals, institutions to contain vested interests, and proper governance of common concerns and resources, tragedy indeed looms. That tragedy is often fanned by narrow-minded, hostile firms, backed by corrupt or at least negligent public authorities. As Bill Nye puts it: "So long as we each focus only on our individual decisions and their short-term consequences, we will act like renters, not owners of this Earth" (Nye 2016, 4). However, as a recent piece in *Nature Climate Change* asserted, there is promise that a network of local institutions, taking a bottom-up approach, can effectively address complex global problems (Vasconcelos, Santos, and Pacheco 2013). There are recent encouraging examples of such progressive coordination.

Initiatives are often taken on the heels of disaster. Hurricane Sandy made New York City and some other U.S. coastal cities reconsider their dependence on climate and ocean; they join Paris, London, Rio de Janeiro, Seoul, Shanghai, and others in the "Compact of Mayors"—a vast movement of cities, also known as C40, the Global Covenant of Mayors, attempting to better govern natural capital. Increasingly severe droughts have also reinforced the determination of Californians to take water management and climate change more seriously; in a not too distant future, they may even convince Texans to switch to the same path. In the Eastern Himalayas, glaciers have already retreated, and river flows have been disturbed to such an extent that alarm is mounting in the countries affected. As the Chinese government expressed in the *Initial National Communication on Climate Change* (2004, 55): "Ice storage in the extremely high mountainous region in China would reduce by a large amount and the capacity of the melted water from glaciers for seasonal regulation of runoff in rivers would lose heavily." China has thus started reconsidering its entire approach to natural capital.

As far as carbon pricing is concerned (more on this fundamental issue in chapter 9), a process of matching initiatives and cooperation in carbon taxation has spread from Sweden to various countries in northwestern Europe. Carbon pricing schemes, admittedly on more modest scales, are also in places in North America and China.

British Columbia and, more recently, Alberta—the province of tar sands—are taking the taxation approach, while California, Quebec, and Ontario have a cap-and-trade system that other Canadian provinces and U.S. states may consider joining.

Financing the transition to a more sustainable development trajectory has not been among the core concerns of capitalism. However, there are signs that this may be changing. In 2014, institutional and private investors with a combined $50 billion in assets pledged at least partial decarbonization of their portfolios; by 2016, the assets in management under this pledge have jumped to more than $3 trillion. Investors are moved to action by concerns for climate change and, even more, concerns for climate change–related shifts on stock exchanges—shifts deriving in part from more stringent public policies, be they actual or expected.

Coal businesses are priority targets in such portfolio rebalancing. They are also priority targets of actions in courts, in political arenas, and on the ground. The Swedish utility Vattenfall has decided to stop producing electricity from lignite (brown coal) and to sell the mines and plants it owns in Germany. Greenpeace (Sweden) intended to buy the business, in order to freeze it and later to close the mines, but it was not allowed to bid—a discrimination that will probably not be considered acceptable in the future. In 2015, Peabody Energy, the largest U.S. coal producer, not only faced competitive pressure from shale gas and the biting regulations of the Obama administration, but also, after a two-year investigation by the New York attorney general, had to acknowledge that it had misrepresented (to investors and regulators alike) the threats that climate change posed to its business and to society at large. By the end of that year, it had lost 93 percent of its value on the New York Stock Exchange; it is now bankrupt. In China, corruption in the coal industry is seriously investigated by the Central Commission for Discipline Inspection of the Chinese Communist Party. Coal was once a sacred cow there; it seems it is no longer.

With coal companies no longer invulnerable, local communities that have long fought against all sorts of pollution now have greater power. In Colombia, for instance, the Wayuu Indian community—the largest in the country—has been at odds with the three largest

Western mining corporations—Anglo American, BHP Billiton and Glencore—over the open pit coal mine Cerrejón, which produces 32 million tons of coal a year. When Yasmin Romero Epiayu, writer and voice of her community, visited the Paris Climate Conference in December 2015, she stressed the dangers of climate change to her community and the disastrous health effects—children mortality rate, in particular—from the mine's pollution. Her arguments were ineffective against the mining companies during the booming years at the turn of the millennium, but the confrontation is getting less unequal. This is important—in many places, from the Arctic to the tropics, confrontations are underway between aliens to the places, which profit from devastating natural capital, and communities that are trying to live in sustainable balance with their natural environment (Descola 2013; Klein 2014).

Everywhere younger generations—and all those to come—shall disproportionately suffer from the damages to natural capital. In growing numbers and in places as diverse as California, Oregon, Texas, the Netherlands, New Zealand, India, and Pakistan, they are suing public authorities for ignoring their legitimate interests in the preservation of natural capital—the climate, in particular. They have often been rebuffed, but they have had some unexpected successes. Public authorities may soon be flooded by complaints and indictments, as passivity, negligence, and gross misconduct become more and more intolerable.

A remarkable case is unfolding in India. The constitution of the country states that it is "the duty of every citizen in India to protect and improve the natural environment, including forests, lakes, rivers and wildlife." Obviously it is no less the government's duty. To say the least, the government doesn't fulfill that duty. The natural environment, far from improving, is degrading at a pace that climate change only accelerates. Ridhima Pandey has filed a petition with the National Green Tribunal—the federal court in charge of environmental affairs—asking the tribunal to order the government "to protect the vital natural resources, on which today children and future generations depend on for survival." Ridhima Pandey is a nine-year-old student living in the state of Uttarakhand, which nowadays is routinely hit by devastating floods and landslides linked to changing conditions in the Himalayan mountains under climate change. Pandey has acted on behalf of fellow students, and more generally of all the children who would suffer in the future. The tribunal has decided to hear the case.

However powerful, all the above stories are still relatively small. Yet they have the potential to grow and coalesce, to force a radical transformation in society and economy. In a different context—life in the Soviet world—albeit with general relevance, Svetlana Alexievich wrote in her 2015 Nobel Lecture: "In my books, people tell their own little histories, and big history is told along the way." As far as climate change is concerned, might big history have been told in Paris at the Climate Conference?

FRAMEWORK, REFERENCE, AND SPRINGBOARD

A mushrooming of new visions and actions converged in Paris, allowing 195 countries to agree on a framework that, if implemented thoroughly, could lead to a serious response to climate change. Whether this new framework will help reverse the trend of catastrophic climate change remains to be seen, but a number of new factors can give some optimism. (We'll go further into COP21 in chapters 10, 11, and 12, where we discuss the geopolitics of environmental issues and sustainability.)

The first factor is the perception of risks. Growing awareness of the magnitude of climate risks has changed the understanding of most governments. Indeed, all major countries have developed their own national scientific assessments of climate impacts. For most of them, the result was frightening and has pushed governments in the face of public opinion pressure to adopt a more proactive attitude on climate policies and the prospect of an international agreement

The second factor, as mentioned earlier in this introduction, has been the increased evidence of existing solutions to the daunting challenge of deep cuts in emissions—plummeting costs of low or zero carbon power, innovation in energy efficiency, and, no less important, the immediate social benefits for public health or traffic congestion of low carbon solutions in domains like energy or transport.

All of this has changed the representation of the cost and benefits of actions. This resulted in an important new feature of the climate regime: the willingness to present voluntary climate plans (national contributions), a bottom-up approach that resulted in a totally unexpected result: 189 countries, covering more than 95 percent of global emissions, presented either before the conference or at the meeting their plans. Even if they fall short from a safe trajectory, these plans represent a serious deviation from "business as usual." All of the plans put forward increased investment in renewable energy, energy efficiency, and forest conservation. Many of these plans are complete and rather detailed; some are still in the making. Nevertheless, it is an unprecedented recognition of the need to change course.

In addition, this massive adhesion to the process had another major benefit. Many research institutes and public institutions evaluated all the plans published in advance of the Paris conference. The assessment was collectively consistent: altogether, the emission reductions embedded in the plans resulted in projected global temperature increase ranging from 2.7°C to 3.2°C above preindustrial levels. The gap between countries' commitments and the global objective has been recognized and accepted by all governments, opening the door for revision mechanisms to ramp up national actions. It was a hard-fought battle to get each government to come back every five years to revise its actions and intensify efforts. The collective acknowledgment of the gap played no small part to win that battle. Beyond these common features, a number of evolutions have contributed to the final result.

The geopolitics of climate talks have been dramatically transformed by the move of the two big players and major emitters—the United States and China. For powerful domestic reasons on both sides, climate change action has to be supported by an international agreement and by parallel and comparable engagement of the other power. In Robert Putman's logic of the "the two-level game," the win-sets of the two countries matched quite well as the proponents of a shift in climate policies. The top levels of the executive branch in both countries—the White House and China's presidency—needed to demonstrate that they were not moving alone (Putnam 1988). This resulted in intensive bilateral discussions held at every level, which led to joint proposals at different points in time. Then the two giants decided they would promote a new growth model—for their countries and for the world—of a progressively decarbonized economy. They agreed on supporting a progressive upgrade of efforts and a common framework of transparency to create confidence in the delivery of actions. Many more useful elements, conceived in the joint statements prepared by the two countries, would finally land in the final agreement.

The key outcome of these discussions was to redefine the north-south divide in the climate regime. From the 1992 UN Framework Convention on Climate Change (UNFCCC) signature onward, the principle of common but differentiated responsibility was a major and intangible principle built into all climate talks. It defined two categories of countries, with clear differences in obligations. As such, it was translated in the Kyoto Agreement by the absence of any constraints on emission reductions from developing countries, whereas developed countries had to undertake economy-wide absolute emission reductions. This divide, a major political obstacle for any climate agreement on the U.S. side, was at the origin of the famous Byrd-Hagel resolution in the U.S. Senate in 1997 (see chapter 9).

Since then, China has become a major competitor of U.S. industry, and the two economies are even more connected. Redefining new concepts of responsibility on greenhouse gas emissions and finance was essential for the United States to back any international framework. China and the United States found together a new balance in which China accepted to participate on a voluntary basis to global climate action in ways similar to developed countries, without eliminating reference to the basic principles of the UNFCCC.

This balance was reflected in the agreement text where, in many places, the notion of collective action prevails without ex ante country categorization. The text invites all countries to do their best and to take more stringent targets over time; it invites developed countries to continue taking the lead on actions, finance, and technology. The classical divide is blurred, acknowledging the catching up of many developing countries and the growing impact of the economic growth of developing countries. In this effort of unlocking political obstacles, the two major emitters did not look for a "separate" bilateral agreement. Instead, they effectively supported the multilateral process, while exercising their influence and soft power on a group of countries providing some of the key elements of the global regime. This was a real shift in many years of U.S. policy (see chapter 10).

Another new political development was the constitution of a block of active countries sharing a common specific target: maintaining the temperature goal well below 2°C and building a system as stringent as possible to force all major emitting countries to cut their emissions in a way consistent with this goal. The "high ambition coalition"—initially regrouping the European Union, small islands states, and some Latin American countries—was later joined by the United States and others, including Brazil, at the Paris conference. Compared to previous rounds of negotiation, this group was larger and more influential, reflecting the global shift among developing countries. The more defensive developing countries lost influence as the prospects for a low carbon economy looked more promising and safer. Expectations that the green and "decarbonized" economy was a foreseeable future, a positive outcome already under way, were increasingly shared between governments. In a climate regime built "bottom up" on voluntary commitments, the best enforcement possible of any agreement is based on a self-fulfilling prophecy. If enough signals converge toward the new trend, every actor will become an active operator and make the prophecy happen. That is the final dimension of the Paris event. Climate change and many other ecological disruptions are the result of multiple decisions at many levels. The number and complexity of drivers make a top-down regime working through a central coordination mechanism almost impossible to achieve. The process in Paris fully recognized this hurdle. As the host of the conference, the French government pushed for the involvement of as many voices as possible, counting on the alignment of objectives among actors and not on a centralized coordination.

The political momentum of Paris thus played a catalyst role for a wide range of institutions and actors. The formal negotiations between governments were embedded in a much broader discussion in which each significant actor or institution wanted to have a say through its own climate policy. Throughout the "regime complex" analyzed by Robert Keohane and David Victor (2010), every institution signaled before the Paris conference its own targets, creating a positive competition.

The close association of nonstate actors to the talks fostered the convergence of expectations. From businesses to cities to influential nongovernmental organization, all networks of climate actors contributed to a new narrative by taking commitments of greenhouse gas emission reductions, announcing targets of investments in renewable energy, and, as mentioned earlier, letting down portfolio assets linked with fossil fuel–based activities. The Paris Agreement has established a platform with shared objectives. The climate cause has gained much ground, but the world is running against time.

1

EROSION OF BIOLOGICAL DIVERSITY

BIODIVERSITY, OR the biological diversity of life forms, is both essential to life on earth and under severe threat. The "erosion" of this biodiversity is actually an understatement. In a not-too-distant future, it may turn into full collapse, as an editorial in *Nature* (Cardinale et al. 2012), motivated by the June 2012 Rio+20 Conference and empirical studies (Newbold et al. 2016), made clear.[1] This chapter discusses how essential biodiversity is and then turns to a discussion of the threats to biodiversity.

Biodiversity provides not only a reservoir of food but also therapeutic and prophylactic resources. Most of what we eat has its origins in biodiversity—for example, wheat in ancient Mesopotamia and corn in Mexico both evolved from wild plants. More recently, a variety of rice widely grown in South Asia was saved from a deadly virus by the introduction into its genome of a gene obtained from an otherwise unremarkable wild variety of rice. In terms of the therapeutic benefits of biodiversity, the most well-known gift of nature is aspirin, which was initially extracted from the bark of the willow tree. More generally, about 50 percent of all drugs approved by the U.S. Food and Drug Administration (FDA) originated from natural substances. Biodiversity also provides natural pesticides.

Biodiversity also plays a fundamental role in the resilience of ecosystems and the sustainability of food chains. As discussed later in this chapter, the eradication of the wolf from Inner Mongolia contributed to turning grassland into sandy desert. Likewise, the decline of bee populations, which is spectacular in some places, deeply disrupts vast ecosystems. Admittedly, some of these ecosystems are "artificialized" agricultural regions, but they are essential for food production. These are only two examples from a long list of ways in which "biodiversity loss might affect the dynamics and functioning of ecosystems" (Cardinale et al. 2012, 59). The reverse—that is, the dependence of biodiversity on healthy ecosystems—is no less true.

However essential in so many respects, biodiversity is under severe multiform threats. One such threat that is particularly detrimental is the destruction of habitats, including oceans, forests, wetlands, and rivers. The various forms of predation (overfishing, deforestation, overgrazing), which is often linked to the destruction of habitats, are also responsible for huge damage. Climate change magnifies these threats (see chapter 4). For the oceans, the current rate of degradation is leading to a general and irreversible impoverishment in the variety of life forms they harbor. For the forests, it leads to terminal decay, most notably in the rain forests.

In this globalized world, we should not underestimate the often devastating effects that inadvertent or voluntary introduction of alien species might have. These species, which are often introduced into ecosystems with a dearth of natural predators, prosper and multiply to the point of severely reducing the preexistent biodiversity.

All in all, the prospects are bleak. As Johan Rockström wrote in the introduction to the *WWF Living Planet Report 2016*: "This single word [*Anthropocene*] encapsulates the fact that human activity now affects Earth's life support system" (World Wildlife Fund 2016, 4). It seems that ever-increasing human activity is increasingly incompatible with safeguarding biodiversity and natural capital in general. Hence, the question: Are we heading (fast) toward a sixth extinction and an inhospitable planet?

DIVERSITY FOR ALL NEEDS

BIODIVERSITY AS A SOURCE OF GENES USEFUL FOR FOOD PRODUCTION

Many well-documented cases illustrate the need for biodiversity as a source of useful genes for food production—from the early domestication of cereals in Mesopotamia and Mexico to the rescue in the 1970s of a variety of rice commonly grown in Asia. In the latter example, the rice had fallen victim to a hitherto inactive virus, with yields reduced by more than 25 percent. This decline was a catastrophe for the farmers, who tend to function not far from some tipping point. No chemical cure could be found. However, in the collections kept by the International Rice Research Institute, located in the Philippines, scientists identified a strain of wild rice that proved resistant to the virus. The strain had been growing in a Philippine valley that had been flooded a few years earlier by the completion of a dam. The variety could not be commercialized, as its yields were too poor. However, scientists were able to transfer the genes responsible for the resistance to the threatened commercial variety (Heal 2000). This is a significant illustration—though by no means isolated—of the importance of keeping biodiversity as broad as possible, if only from an insurance perspective.

BIODIVERSITY AS A SOURCE OF THERAPEUTIC AND PROPHYLACTIC COMPOUNDS

In terms of the therapeutic and prophylactic benefits of biodiversity, there are numerous significant cases, from aspirin extracted from the willow tree at the end of the nineteenth century to paclitaxel (Taxol) extracted from the Pacific yew more recently. In the early 1970s, scientists at the U.S. National Institutes of Health (NIH), while screening a large number of plants for possible therapeutic properties, identified a promising compound in the bark of the Pacific yew. Paclitaxel has since played an essential role as a drug against various forms of cancer—in particular, breast, ovarian, and lung cancers. The demand for this drug quickly became so high that the yew was in danger of extinction, as removing the bark kills the tree. Fortunately, in this case, biochemistry came to the rescue of ecology, when scientists at the Institut de Chimie des Substances Naturelles, near Paris, and then at Florida State University succeeded in semi-synthetizing the compound from the tree's needles.

Paclitaxel acts by inhibiting cell division; unfortunately, its action is not restricted to cancerous cells. However, derivatives of combretastatin—a compound identified in the bark of the combretum, or bush willow (a tree common in South Africa)—have been found to be more specific. Scientists and researchers have synthesized various derivatives of the compound that have remarkable properties: they act on cells as paclitaxel does, but they can be switched on and off by light. On their path within the body, they can thus be activated by illumination only when required. These results—a joint product of biodiversity and advanced biochemistry—have been reported in *Cell* (Borowiak et al. 2015).

There is a tree in India—subsequently introduced in Africa and elsewhere around the world—called *neem* that has been of great value to Indian farmers for millennia. The juice in its fruits contains a natural pesticide with two remarkable properties: it has a large spectrum, and it is biodegradable shortly after application. It is a perfect example of a gift of biodiversity recognized in traditional knowledge. However, in the 1980s, U.S. chemical firm W. R. Grace became interested in the potential of the neem fruit. After having marginally improved the quality of the pesticide made from the fruit, W. R. Grace applied for a patent at the U.S. Patent and Trademark Office (USPTO) and subsequently at the European Patent Office (EPO). As too often happens at patent offices (see chapter 8), especially at USPTO, the patents granted were too broad, covering not only the upgraded product but also *any* product extracted from the neem fruit. This was a manifest abuse of intellectual property. In essence, Indians were banned (for the twenty-year duration of the patent's life) from selling any product derived from the neem fruit on U.S. and European markets. That was, of course, unfair, though it was more symbolic than effective, as the Indian farmers had no intention of starting such a trade. Rather effective, however, was the decision by W. R. Grace—once it had secured monopoly power in the United States and Europe on the basis of its broad patents—to buy large quantities of neem fruit from India. In response, prices shot up, depriving many Indian farmers of a traditional and essential input. The neem case has been an incentive to include effective benefit-sharing mechanisms in the Nagoya Protocol, a supplementary agreement to the UN Convention on Biological Diversity (CBD) that entered into legal force in October 2014.

In addition to the essential pharmaceutical and prophylactic resources provided by biodiversity, another two have rather unexpected uses. When the DNA of a criminal is found on a victim, it is often in very small quantities—too small to provide valuable information. It is thus necessary to start and sustain a biochemical process to produce replicas of the original DNA. This process was made possible by the introduction of enzymes able to survive at very high temperatures. Such enzymes have been discovered in the waters of the geysers of Yellowstone National Park. The second unexpected use involves a compound utilized in hydraulic fracturing ("fracking"). Guar, a bean harvested mainly in the semiarid region of Jodhpur (in Rajasthan, India), is traditionally used as a foodstuff, but a gum made from guar beans is also used to increase the viscosity of fluids injected into the fracking wells. With the recent and explosive development of gas and oil extraction by fracking (see chapter 3), the demand for guar shot up, as did the price. According to Halliburton, the giant oil and gas services firm, in certain basins, guar gum accounts for up to 30 percent of registered fracking costs.[2] In fact, some operations have had to be delayed due to a lack of guar gum.

As is clear, there are significant economic and social effects of what are sometimes unremarkable elements of biodiversity. Trucost, an environmental data provider, estimated that, in 2010, the eight hundred largest companies in the world, from almost all sectors of activity, extracted from free biodiversity and ecosystem services no less than $850 billion in benefits (Trucost.com 2011). This is on top of the goods and services (foodstuffs, wood, inputs for drugs and perfumes, specific industrial inputs) that these companies procured from markets.

BIODIVERSITY AND THE SUSTAINABILITY OF ECOSYSTEMS

At one time on the Mongolian plains, the wolf fed on the antelope, the antelope fed on the grass, and the grass fixed the soil. The Chinese authorities in Beijing, seeing the wolf as an intrinsically dangerous and useless animal, decided it should be eradicated—and so it was. The antelopes proliferated and overgrazed. The grass disappeared. And then the antelope followed suit. Now, only bare sandy soil remains (Rong 2009). Each spring, strong winds bring increasing quantities of that sand to Beijing and even farther away.[3]

BOX 1.1 VULTURES' DOOM

Vultures have a poor reputation in both their real and symbolic lives. They are almost never heroes in children's books. A brilliant exception, however, is Tomi Ungerer's *Orlando*, the story of a brave Mexican vulture who rescues an American family in the desert and befriends local Indians. Ungerer has a point: vultures protect people and ecosystems, but they are poorly rewarded for their services.

Because vultures are meticulous scavengers, they are efficient cleaners in the wild. Moreover, they are sanitation agents, as specific acids in their digestive juices destroy pathogens that proliferate on dead bodies. Yet, their populations in Africa and Asia are fast dwindling:

- It is more and more difficult for them to find convenient habitats.
- The use of parts of their bodies in traditional medicines is not abating.
- They are poisoned accidentally—for example, by drugs administered to livestock that are lethal to the birds.
- They are poisoned on purpose or killed in other ways—for example, poachers of elephant ivory or rhinoceros horns kill vultures circling over dead animals as this can alert gamekeepers.

To make things worse, the species is especially vulnerable due to vultures' low reproduction rate.

The content of this box is based on Charlie Hamilton James's exhibits at the Wildlife Photographer of the Year exhibition at the Museum of National History in London (October 2016 to September 2017).

It is well known that many fish species depend on coral reefs as their home ecosystem. The reverse is also observed, as testified by the essential role played by parrotfish in Caribbean coral reefs. Parrotfish are grazers that check the growth of algae that live in a symbiotic relationship with the corals. They also eat sponges, the growth of which would otherwise choke the corals. However, parrotfish in the Caribbean have been overfished or destroyed as bycatch, both of which appear to have significantly contributed to the degradation of most reefs in the region. Indeed, places where parrotfish have been protected—in particular, Bermuda and Bonaire—have the healthiest corals. In the waters around Barbuda, where strict protection measures were introduced in 2012 as recommended by a joint report from the Global Coral Reef Monitoring Network, the International Union for Conservation of Nature (IUCN), and the UN Environment Programme (2012), positive effects on coral were swiftly observed.

In Australian woodlands in the state of New South Wales, two ecologists from Charles Sturt University performed an experiment showing that mistletoe is good for more than providing kissing opportunities on Christmas Day. Indeed, this plant appears to be an ecosystem "keystone resource" that makes tree canopies more attractive for certain species of birds (Watson and Herring 2012)

According to the Natural Resources Conservation Service of the U.S. Department of Agriculture, three-fourths of the world's flowering plants and about 35 percent of the world's food crops depend on animal pollinators to reproduce (NRCS 2017). The *Assessment Report on Pollinators, Pollination, and Food Production,* produced by the Intergovernmental Science-Policy Platform on Biodiversity and Ecosystem Services (2017, 18), also stresses the importance of animal pollinators: "In the absence of animal pollination, changes in global crop supplies could increase prices to consumers and reduce profits to producers, resulting in a potential annual net loss of economic welfare of $160 billion–$191 billion globally to crop consumers and producers and a further $207 billion–$497 billion to producers and consumers in other, non-crop markets (e.g., non-crop agriculture, forestry, and food processing)." The report thus expresses deep concern about the severe declines observed in the populations of animal pollinators.

Bees play a prominent, albeit far from exclusive, role in pollination. During the past fifteen years, beekeepers have noticed a severe decline in bee populations, including sudden implosions of entire colonies.[4] During the past five years, that decline has turned to collapse in key rural regions in the United States and Europe. The causes of the colony collapse phenomenon have proved difficult to investigate scientifically. Two studies reported in *Science* (Stokstad 2012) have significantly enhanced the credibility of the following explanation: although general factors, such as the impoverishment of ecosystems in rural regions, are highly detrimental to bees, there are also specific factors that better correspond to the brutality of the phenomenon. The scientific results mainly point to the extensive use of pesticides in industrial agriculture—in particular, a class of pesticides called *neonicotinoids*. These pesticides are not particularly toxic to mammals, which was a good argument for them to be authorized by regulators; however, this argument neglects the fact that neonicotinoids are highly toxic to all sorts of insects. Even when they are not directly lethal, they act on the nervous system of insects in ways that indirectly provoke individual and collective deaths. For example, sublethal doses of neonicotinoids damage bees' memories, their ability to forage, and their ability to navigate back to their hives. This pesticide also drastically inhibits the production of queens and weakens bees' defenses against the damaging effects of fungi, parasites like mites, and viruses.

It is striking that while bees struggle in the rich agricultural regions around Paris, they thrive in hives on the roof of the Paris Opera, in the gardens of Sciences Po university at St.-Germain-des-Prés, and in municipal parks in Paris—all of which are effected by the ban on chemical treatments within Paris. In the United States, the distribution of colony loss is heavily skewed toward states where the use of neonicotinoids is at a maximum: beekeepers lost more than 60 percent of their colonies from April 2014 to April 2015 in Illinois, Iowa, Oklahoma, and Wisconsin, compared with 30 percent in northwestern states (Bee Informed Partnership 2015).

Damages caused by neonicotinoids are by no means confined to honeybees, as has convincingly been shown by investigations of the European Academies Science Advisory Council (2015) in a report sponsored by nineteen European academies of science: "There is an increasing body of evidence that the widespread prophylactic use of neonicotinoids has severe negative effects on non-target organisms that provide ecosystem services including pollination and natural pest control" (29). A British study spanning more than eighteen years confirms this assessment (Woodcock et al. 2016), as do repeated surveys in German nature reserves (Vogel 2017).

The situation is mind-boggling: pollinators act as mediators in ecosystems whose sustainability depends upon this very mediation. The economic values involved are vastly superior to the profits made from producing and selling the offending pesticides. Yet, neither market nor regulatory mechanisms have corrected these "imbalances." Although an international regulation institution is nominally in charge, it is controlled by the pesticide oligopoly, making it a classic example of regulatory capture and regulation impotence.

The wolf and the bee play specific roles in their respective ecosystems. They also illustrate a general trend emphasized in the aforementioned article published in *Nature*. Cardinale and his colleagues derived six scientific consensus statements from the body of research produced during the past twenty years. The first consensus statement underscores the contributions of biodiversity to the sustainability of ecosystems: "There is now unequivocal evidence that biodiversity loss reduces the efficiency by which ecological communities capture biologically essential resources, produce biomass, decompose and recycle biologically essential nutrients" (Cardinale et al. 2012, 60). The concern expressed here is obviously of general relevance.

DESTRUCTION OF HABITATS AND PREDATION

RAIN FORESTS UNDER SIEGE

Among terrestrial ecosystems, rain forests are the richest reservoirs of biodiversity and the most efficient stabilizers of climate. In fact, in this latter role, they are matched only by oceans. Let's consider the transformations they have undergone for the past fifty or sixty years, starting with the Amazon rain forest, which comprises more than half of the total remaining area of rain forests worldwide. The Amazon rain forest covers approximately 5 million km^2 of the Amazon

River basin, extending across nine countries. Brazil is dominant, holding 60 percent of the forest area. The Amazon is home to the greatest variety of plants and animals on earth. It has been estimated that 1 km^2 contains up to several thousand types of trees and other higher plants. More than 100,000 different invertebrates have been described, and that number may be less than 10 percent of the actual number. Birds are also well represented, with 20 percent of all existing birds worldwide living in the Amazon.

Until the 1960s, the interior of the Amazon was barely accessible and remained almost untouched. But then conditions radically changed with the construction of roads penetrating deep into the forest and the clearance of large tracts for cattle ranching and soybean farming. Deforestation peaked in 2004, with about 27,000 km^2 cleared. A stricter implementation of the Brazilian Forest Code helped reduce the clearances to 6,000–7,000 km^2 in 2011. However, that enforcement proved unacceptable to the powerful agricultural lobby (the main Brazilian export sector, in addition to the mining sector), which accused environmentalists of working hand in hand with foreign interests and putting at risk the competitiveness of the Brazilian economy. The lobby persuaded Brazil's lower house of Congress to vote for a bill stripping the Forest Code of essential provisions for protection of the forest. After the Senate declined to make significant modifications, the bill was submitted to President Dilma Rousseff. The changes that the president brought to the bill are not negligible, but they are proving insufficient to prevent a reversal of the decreasing trend observed between 2004 and 2011. Nor do they stop the increase in criminal activities throughout the forest, with long-established indigenous communities targeted in particular. In addition, the strong protections of zones that have particularly rich biodiversity, such as riverbanks, mangrove swamps, and mounds, have not been reintroduced.

Coupled with the effects of climate change, an upsurge in clearances might tip the Amazon into overall and irreversible decay.[5] According to an article in *The Economist* (2010): "If the Amazon went up in smoke—a scenario which a bit more clearance and a bit more warming makes conceivable—it would spew out more than a decade's worth of fossil fuel emissions." In fact, "a decade's worth" might be an underestimation, as feedback effects are certainly at work. As forest cover regresses, it is getting drier in the dry season, and wildfires are becoming more frequent and more devastating. In some years, the areas affected by wildfires have been of the same order of magnitude as those deliberately burnt or cut down. Not only is the carbon contained in trees released into the atmosphere, but carbon fixed in forest soils is released as well. Microorganisms normally active only in the topsoil are activated by warmer conditions linked to deforestation to the point of being able to digest carbon fixed deeper in the soil, producing CO_2 in the process. The results from systematic observations on 321 plots in the interior of the Amazon rain forest over three decades do not provide encouraging perspectives. These findings have been published in *Nature* (Brienen et al. 2015) under a transparent title: "Long-Term Decline of the Amazon Carbon Sink."

Another rain forest in Latin America—the Paraguayan Chaco—is particularly rich in biodiversity, despite its much smaller size (300,000 km^2). Since 2006/2007, it has been destroyed by local and Brazilian cattle ranchers at a significantly faster rate—2 percent per year—than the Amazon, without attracting much attention.

The second largest rain forest, and second richest in biodiversity, is in Central Africa, in the Congo River basin. Here, destruction does not stem from inadequate laws or imperfect implementation. Instead, this is one of the most desperately lawless regions on earth, devastated by war and inept (or nonexistent) governance. Logging, slashing, and burning proceed chaotically, benefiting warlords, corrupt politicians and officials, and fraudulent foreign businesses. International institutions like the World Bank have tried to bring in some remedies, though they acknowledge their failure.

The third significant group of rain forests is in Southeast Asia, extending from India to Malaysia, Indonesia, and Papua New Guinea. These forests are exploited and burnt to the point at which their very existence hangs in the balance[6]—not to mention the thick smoke polluting the region. In most places, public authorities are no match for international logging companies that sell wood all over the world—though mainly in China, Japan, and India; for paper mills, such as the powerful Indonesian group Indah Kiat Corporation; or for agribusinesses that convert natural forest into palm oil plantations. Indeed, many public authorities are generously rewarded accomplices in this destruction. However, the ultimate culprits are the U.S., Asian, and European processed food companies and their customers, along with the numerous banks that finance the logging, production, and distribution processes involved (RAN 2016).

Although this may look more "civilized" than what is going on in the Congo basin, the consequences are worse in terms of forest and biodiversity losses. If the present trends are maintained (and there are not many credible counterweights), most of the Southeast Asia rain forests will be gone by midcentury—and with them a rich and original biodiversity.

For the Amazon and the Congo rain forests, it might take more time, but doom appears to be no less inevitable, unless the present trends are decisively reversed. But this reversal will not be easy to achieve, as it involves confronting powerfully organized, narrowly—albeit highly—motivated—economic and political forces that don't shy away from using manipulation, corruption, intimidation, and outright violence. Although there has been much talk of buyout mechanisms, the money has not been very forthcoming (see, however, chapter 6 for an interesting exception).

OCEANS' DOOMSDAY

In April 2011, an international scientific workshop convened under the auspices of the International Programme on the State of the Ocean and IUCN set out to assess the latest information about the condition of the oceans. They summarized their findings as follows:

> The participants concluded that not only are we experiencing severe declines in many species to the point of commercial extinction in some cases, and an unparalleled rate of regional extinctions of habitat types (e.g., mangroves and seagrass meadows), but we now face losing marine species and entire ecosystems, such as coral reefs, within a single generation. Unless action is taken now, the consequences of our activities are at a high risk of causing, through the combined effects of climate change, overexploitation, pollution and habitat loss, the next globally significant extinction event in the

> ocean. It is notable that the occurrence of multiple high intensity stressors has been the pre-requisite for all the five extinction events of the past 600 million years. (Rogers and Laffoley 2011, 7)

Thus, the perspective is no longer of biodiversity erosion but of biodiversity collapse.

Chapter 4 addresses the effects of climate change on the oceans. Let us here consider overexploitation of the oceans, which results from the conjunction of (essentially detrimental) technical innovation and the "free-rider effect" plaguing public goods (as analyzed in detail for fisheries in Sumaila 2013). It also results from plain criminal activities and is boosted by perverse public subsidies.

Present fishing technology includes fish-aggregating devices, trawls, lines with thousands of hooks, dredges, echo sounders, computers, and satellite links, to name only a few. The power of such technology to catch fish and destroy habitats has grown out of all proportion to the regeneration capacities of populations and ecosystems.[7] There are neither effective rules nor institutions to regulate and monitor the high seas—those waters that are beyond the 200 nautical miles (370 km) that define the limits of the national exclusive fishing zones. In principle, the high seas are protected by the 1982 UN Convention on the Law of the Sea; in practice, however, the convention is toothless. In fact, the United States didn't even bother ratifying it. Over the high seas, the logic of the free-rider effect is implacable—that is, "What I don't catch or destroy, someone else will catch or destroy."

South Atlantic waters off the coast of western Africa offer a striking example of pillage by foreign fleets (mainly from Asia), with ships equipped with the most efficient catching devices, not to mention the support of their respective governments. Among these ships are those of the state-owned China National Fisheries Corporation, the exactions of which are not in tune with the spirit of cooperation that China is supposed to display in Africa.

Public subsidies magnify the problems. Evaluating such subsidies on the basis of a methodological note prepared by the Directorate-General for Internal Policies of the European Parliament (Sumaila et al. 2013) and of a World Bank Report (World Bank 2017), annual global public subsidies to fisheries are on the order of $50 billion. Such subsidies are particularly perverse when they pay to enhance the capacity of fishing fleets, which is what more than half the total amount of subsidies does.

In *It Happened Tomorrow,* an American film directed by René Clair set during World War II, the hero comes across a copy of *Tomorrow Daily*, in which he reads an account of his own death. In the present desperate situation of a few fishing communities in India and the Philippines, thousands of other fishing communities around the world can read their own future demise. According to Jim Yardley, in Vellapallam, a village in the Indian state of Tamil Nadu, local fishermen face a stark choice: stop fishing and starve, as there is no other possible trade, or go on fishing and risk being abducted or killed on the spot by the Sri Lankan Navy, as has already happened to more than a hundred fishermen (*New York Times*, September 5, 2012). Indeed, since the 2004 tsunami, modern trawlers financed from relief funds and docked at nearby towns have depleted the fish stocks in Indian waters. In addition, since the end of the Sri Lankan civil war in 2009, the Sri Lankan Navy has returned to police the straights between India and Sri Lanka to protect Sri Lanka's "natural resources."[8]

Around the islands of Cebu and Palawan in the Philippines, the only remaining fish are in reefs at significant depths (down to 40 meters). The fishermen practice a method of fishing called *pa-aling,* which is one of the most exhausting and dangerous techniques of all: fishermen equipped with crude, unreliable respiratory devices dive with a net to trap the fishes from the deep reef.[9]

Along the coasts of Madagascar, among the coral reefs off Mozambique, on the shores of Lake Tanganyika, and in many other places of southern Africa, another *It Happened Tomorrow* scenario is unfolding. Bloody wars are being waged between poor local professional fishermen and desperate landless families who catch scores of juveniles and other creatures that proper fishing nets would spare, thus depleting the stock of fish. And these families catch the fish with mosquito nets provided to them as protection against catching malaria.

Pollution under various forms also contributes to the oceans' decay. Dead zones result from excess nutrients deposited by rivers that flow through intense industrial agricultural regions. The Mississippi River, for example, gathers nitrates and phosphates in the Midwest and carries them to the Gulf of Mexico, leading to algae bloom and then decay, sucking all the oxygen from the water. Dr. Melissa Baustian, a coastal ecologist at Louisiana State University, used to dive in the dead zone: "The deeper we go down in the water, it gets kind of scary, because there is nothing there. There's no fish, there's no organisms alive—so it's just us." At its peak in summer, the Mississippi dead zone extends more than 20,000 km^2. The dead zone off the coast of northern Brittany (France) is much smaller, but the degradation is also severe. The plague is also catching up in coastal zones of China and Japan, as well as in the Bay of Bengal, where the survival of tens of millions of people who make a living from the sea is at stake.

Another factor contributing to the decline of the oceans is the concentration of plastics that marine currents bring to remote places. In one such place, an armada of about 12,000 plastic ducks (of the sort children play with in the bathtub) was spotted (Ebbersmeyer and Scigliano 2009). More seriously, because many birds cannot distinguish between plastic and flesh, they are starving their chicks by feeding them plastic items. Similarly, fish are absorbing increasing quantities of those sub-millimeter-sized plastic granules that cosmetics producers integrate into hand lotions and face creams as exfoliants.

In addition to the well-known zones of convergence and accumulation in the North Pacific (e.g., the "Great Pacific Garbage Patch") and the North Atlantic, which are growing at an alarming rate, other critical zones have been identified in the southwestern margins of the Indian Ocean and in the Pacific Ocean. Of particular note is the Tasman Sea, where the damage is magnified by the coincidences of high concentrations of debris and an unusually large diversity of seabirds. In addition to birds feeding plastics to their chicks, they are also suffocated by getting trapped in plastics or even poisoned. Poisons from certain components of the plastics may leak inside their bodies after being ingested; there is even evidence that poisons are transmitted to the progeny.

According to the broad inquiry reported in Wilcox, van Sebille, and Hardesty (2015), "Concentrations [of plastics] reach 580,000 objects per km^2, and plastic production is increasing exponentially" as confirmed in Geyer, Jambeck, and Law (2017). It has been shown that 59 percent of the seabird species studied between 1962 and 2012 have ingested plastics; at the current rate of increase, 99 percent will have done so by 2050. However, this trend is not fatal. Indeed, significant decreases in the accumulation and ingestion of debris have been observed in the North Atlantic and are linked to systematically sounder management of the production, use, and disposal of plastics in northern European countries.

Taking into account the residue of the innumerable products created by the chemical and pharmaceutical industries, will we soon have oceans empty of fishes and corals but full of pollution and jellyfish, the latter of which prove remarkably resilient? It is quite possible. However, the vitality with which diversity of life comes back in protected zones (provided a no-return situation has not been reached) points to a possible different direction (see chapter 6).

DISSEMINATION OF ALIEN SPECIES

Although the dissemination of alien species is a less conspicuous factor of biodiversity regression, it is a very significant one when those species become invasive in ecosystems where they face no check on their development. The globalization of trade and tourism has multiplied the number of occurrences—and the effects are indeed global. As Anthony Ricciardi forcefully put it: "We're talking about the widespread redistribution of life on earth—there is no precedent in the fossil record" (quoted in Davison and Burn-Murdoch 2016).[10]

Let's look at a couple of illustrative cases. The first one might look parochial, but it is representative of many flawed voluntary introductions. An Asian variety of ladybug, called the harlequin, which is rather bigger and more voracious than the European varieties, was introduced in the Netherlands and southwest England as a more efficient vector of biological warfare against various pests. In reality, however, harlequins have proved much better at eating the indigenous varieties of ladybugs than the targeted pests. They have proliferated to such an extent that they have become pests themselves.

Jellyfish from the coast of Maine have inadvertently been brought to the Black Sea. Higher water temperatures and plentiful food in the Black Sea provoked an explosion of their population. These jellyfish have eliminated several indigenous species—in particular anchovies, which were an important catch in the region. This is just one illustration of the general advance of jellyfish in the oceans. The invasion of Guam by the brown tree snake from Australia and Indonesia might have caught Alfred Hitchcock's attention. Accidentally brought to Guam in the 1950s, the snake found abundant and easy-to-catch prey and faced no predators. It doubled in size and proliferated, reaching densities of up to a hundred snakes per hectare. Guam native vertebrate species have been wiped out, and humans have suffered considerable inconveniences, such as frequent power outages, not to mention sustained stress. Because Guam is a military and trade hub, these snakes have since been carried to Diego Garcia, Hawaii, Okinawa, and even Texas.

Most worrying is the dispersion all over the world of pathogenic bacteria and fungi. An extensive survey published in *Nature* (Fisher et al. 2012) leads to a bleak assessment: "The past two decades have seen an increasing number of virulent infectious diseases in natural populations and managed landscapes. In both animals and plants, an unprecedented number of fungal and fungal-like diseases have recently caused some of the most severe die-offs ever witnessed in wild species" (2012, 186). Among animals, amphibians have paid a particularly high price. The skin-infecting fungus *Batrachochytrium dendrobatidis* has caused many populations to collapse, especially in Central America and Australia (Alroy 2015). In addition, a fungus can act in association with an insect. As one example, the association of the mountain pine beetle and the blue-stain fungus, stimulated by increasing temperatures, is decimating mountain forests in northwestern states and provinces of the United States and Canada (see chapter 4).

It is also worth considering a number of other significant cases that reinforce the assertion that the consequences of the migration of species (i.e., their globalization) should by no means be underestimated.

- The globalization of mosquitos like *Aedes aegypti* and *Aedes albopictus*, which are vectors of dengue and yellow fevers, induces the globalization of these ailments.
- The migration of the bacterium *Xylella fastidiosa* from Californian vineyards has led to the infection of citrus groves in Brazil and olive groves in southern Italy. It has since jumped to Corsica and the Balearic Islands.
- Coming from North America, *Ceratocystis platani,* a fungus and a serial killer of plane trees, obstructs the tree's vascular system. It has been spreading through southern Europe and is now moving north toward cities like Paris and London. There is no known cure, though resistant hybrids have been obtained by crossing American and Oriental varieties.
- From Southeast Asia, the red palm weevil, a reddish-brown beetle, has been advancing westward since the mid-1980s, spreading through many areas on its way to Egypt. From Egypt, it leapt to Spain, and from there invaded southern Europe. Its larvae burrow to feed on the central core of palm trees. The damages are obviously far more significant in Egypt than in France, though the loss of the lines of palm trees along the Promenade de la Croisette in Cannes and the Promenade des Anglais in Nice might legitimately be regretted. Something can, however, be done either by injecting an insecticide into the trunk or by confronting the larvae with a microscopic ringworm that kills them. Once again, biodiversity is here to help.

We have also seen a proliferation of other fungi that affect trees around the world:

- *Phytophthora ramorum*, which badly infects oaks and larches
- *Cryphonectria parasitica*, also called chestnut blight as chestnuts are their main victims
- *Ophiostoma ulmi*, spread by bark beetles among elm trees (Dutch elm disease)
- *Hymenoscyphus fraxineus*, which causes devastating ash dieback

And then there are various invertebrates, like the oak processionary moth. These signs all appear to point to a siege on the trees of Europe. (For trees in the United States and elsewhere, see the earlier discussion in this section.)

On July 13, 2016, the European Commission published a first list of thirty-seven invasive species to be eradicated all over the EU. This decision is justified in the following terms: "There are currently over 12,000 plants, animals, fungi and micro-organisms in the EU that are alien to their natural environment. Some 15 percent of these species are invasive, and their numbers are rapidly growing. By crowding out indigenous species, these invasive alien species are one of the biggest causes of biodiversity loss and have major economic consequences" (European Commission 2016, 1).

TOWARD A SIXTH EXTINCTION?

The information about the state of nonhuman life on earth provided in the *WWF Living Planet Report 2016* is a cause for concern: "Wildlife populations have already shown a concerning decline, on average by 58 percent since 1970, likely to reach 67 percent by the end of the decade" (World Wildlife Fund 2016, 6). Pollution and destruction of habitats are major causal factors for this decline. Freshwater habitats—lakes, rivers, wetlands—have suffered most, and for populations depending on those habitats, there is a stunning 81 percent decline. It seems that frogs will soon be confined to books and pictures.

There are also other factors at work in this decline: the explosion of pest populations and diseases, the dissemination of overpowering alien species, industrialized agriculture, predation by humans, and climate change. Predation takes many forms—from massive overfishing and deforestation to targeting emblematic animals for profit, such as poaching African elephants for ivory, rhinos for the supposed virtues of their horns, and snow leopards for furs and bones, not to mention killing lions "just for fun." These populations are being reduced at an alarming pace. What will our children say after we let all the lions die? (Answer: they will be made to believe that lions—as well as giraffes, elephants, rhinos, tigers, etc.—have been created by a god called Disney.)

Several factors are also at work in the oceans, where destruction of habitats, pollution, predation, warming, and acidification resulting from climate change (see chapter 4) compound the effects. Along with coral reefs and shells, large-bodied animals are particularly at risk. Using a database of 2,497 marine vertebrates and mollusks, Payne et al. found that "extinction threat in the modern oceans is strongly associated with large body size, whereas past extinction events were either nonselective or preferentially removed smaller-bodied taxa" (2016, 1284). They concluded that, given "the differential importance of large-bodied animals to ecosystem function, [this trend] portends greater future ecological disruption than that caused by similar levels of taxonomic loss in past mass extinction events" (1284).

When the number of individuals in a specific species falls below a critical threshold, the species faces extinction. When this is the case for a large number of species at one time, mass extinction looms. We define *mass extinction* as is usual in paleontology: "Palaeontologists characterize mass extinctions as times when the Earth loses more than three-quarters of its species in a geologically short interval, as has happened only five times in the past 540 million years or so" (Barnosky et al. 2011, 51). Are we facing the sixth mass extinction right now?[11] The assessment provided in Ceballos et al. (2015) is almost unanimously shared by scientists working in the field:

> The evidence is incontrovertible that recent extinction rates are unprecedented in human history and highly unusual in Earth's history. Our analysis emphasizes that our global society has started to destroy species of other organisms at an accelerating rate, initiating a mass extinction episode unparalleled for 65 million years. If the currently elevated extinction pace is allowed to continue, humans will soon (in as little as three human lifetimes) be deprived of many biodiversity benefits. On human time scales, this loss would be effectively permanent because in the aftermath of past mass extinctions, the living world took hundreds of thousands to millions of years to rediversify. (4)

Three human lifetimes is indeed a rather short geological time period for such momentous transformations.

2

THE UBIQUITOUS WASTE AND GROWING SCARCITY OF WATER AND SOIL

WATER IS essential not only for survival but also for the bare necessities of life. It is also a necessity for the realization of each individual's potential, as illustrated by the testimony of a ten-year-old girl from El Alto, Bolivia: "Of course I wish I were at school. I want to learn to read and write. But how can I? My mother needs me to get water" (UN Development Programme [UNDP] 2006, 1). Although the United Nations has proclaimed water a human right, we are facing a dilemma, because water is not a public good that may be enjoyed collectively, and in most situations, water availability is limited.

A great deal of water is leaking along the networks that serve households in both developed and less-developed countries. Industrial plants in emerging and developing countries tend to waste and pollute water significantly more than those in developed countries. However, around the world, much more water is wasted quantitatively in agriculture than from anything else. Irrigation is responsible for slightly more than two-thirds of the total quantity of water consumed on earth, and about half of that amount is lost.

It is obvious that the present situation is unsustainable. However, the potential for improvement is proportional to the size of the losses.

Drastically innovative methods—both technical (e.g., drip irrigation) and managerial (e.g., transferable water rights)—have been designed and implemented from Israel to Nebraska, from Jordan to Indonesia, from Oman to Australia (see chapter 6).

Properly managing water resources within national boundaries is tricky enough. The difficulties are magnified when a significant resource, such as a main river or a large aquifer, is shared among several countries. In 1863, Mark Twain, then a reporter at the Nevada *Territorial Enterprise*, was said to have uttered: "Whiskey is for drinking; water is for fighting over." That was in the nineteenth-century setting of the Colorado River; it unfortunately remains relevant—at least as far as water is concerned—in the twenty-first-century context of rivers as diverse as the Nile, the Mekong, and the Euphrates-Tigris system, as well as of aquifers like the North-Western Sahara Aquifer System and the Indus River plain aquifer.

Most rivers and aquifers (not only the international ones) also suffer from the free-rider effect—that is, disregarding the effects of one's action (e.g., pumping water) on the resources available to others and on the pace of resource depletion. Yet, cooperation can be made possible, as evidenced by the Guarani Aquifer Agreement between Argentina, Brazil, Paraguay, and Uruguay.

In 2010, *Nature* invited several scientists to identify the main issues in their respective subjects as they looked toward the 2020 horizon. David R. Montgomery, professor of earth and space sciences at the University of Washington and author of *Dirt: The Erosion of Civilizations* (2012), candidly assessed the future of soils:

> To avoid the mistakes of past societies, as 2020 approaches, the world must address global soil degradation, one of this century's most insidious and underacknowledged challenges. Humanity has already degraded or eroded the topsoil off more than a third of all arable land. We continue to lose farmland at about 0.5 percent a year—yet expect to feed more than 9 billion people later this century. . . . Ensuring future food security and environmental protection will require thoughtfully tailoring farming practices to the soils of individual landscapes and farms, rather than continuing to rely on erosive practices and fertilizer from a bag. (Montgomery 2010, 31)

Soil is an immensely diversified and rich—albeit, fragile—ecosystem, as this chapter explains in some necessary detail. Yet, conventional industrial agriculture does not rely on ecosystemic diversity and richness, instead choosing to ignore its agricultural potential. To those in the industry, soil is a mere carrier of chemical implements transformed into food with the help of photosynthesis. Record, though unsustainable, yields have been obtained in this way, as have record levels of water, air, and soil pollution. This is the approach to agriculture that land-grabbers, adding insult to injury, are implementing in Africa and in poor Asian countries like Cambodia and Laos. Latin America has more recently also been of concern. All of this provides a textbook case of unsustainability.

Alternatives with a more sustainable future are investigated in chapters 6 and 8.

WATER: A PRIVATE GOOD THAT IS ALSO A HUMAN RIGHT

A STICKY DILEMMA

In 2010, the UN General Assembly recognized that water and sanitation should be human rights, as access to safe, affordable water is a pillar of economic and social development. Water is essential not only for biological survival (think of the children with nothing but wastewater to drink) but also as a condition for the realization of each individual's potential.[1] There is, however, a tension between the recognition of water as a human right and the indication that it should be affordable. What is the affordability of a human right? Would affordable mean free of charge?

For President Julius Nyerere in Tanzania, the answer to this last question was yes. He acted accordingly, though with consequences he didn't expect:

> Growing up on the shores of Lake Victoria in the 1950s, Anna Tibaijuka would earn a couple of cents by sorting coffee beans for her father. With one of these coins, she would buy a sweet from an

Indian shopkeeper. With the other, she would buy potable water from a kiosk.

But when she returned to her hometown in the early 1960s, the kiosk was no more. Julius Nyerere, Tanzania's first president, had declared water free. When it cost a cent, not a drop was wasted, Mrs. Tibaijuka recalls. But when the tap ran freely, water was squandered, and—inevitably—stopped. (*The Economist*, 2006)[2]

The dilemma may be discussed in simple economic terms. Potable water is indeed an essential good; however, it is also a private good, the use of which is exclusive (any quantity consumed by A cannot also be consumed by B). Thus, it is not a public good. As soon as water availability is limited, the very fact that water is a private good, and hence must be shared, is incompatible with free access.

UPSIDE-DOWN PRICING

Does limited availability combined with being a private good imply that poor people are excluded? They often are. They then try to circumvent that exclusion by having women and children haul water long distances or by consuming unsafe water. If they decide to avoid both of those situations, they often end up paying for their water at much higher prices than people who are not poor. According to the UNDP (2006, 2), "In high-income areas of cities in Asia, Latin America, and Sub-Saharan Africa, people enjoy access to several hundred liters of water a day delivered into their homes at low prices by utilities. Meanwhile slum dwellers have access to less than twenty liters of water a day per person and pay per unit of water, delivered by tank, five to ten (and in extreme cases up to forty) times more than the inhabitants of richer areas."

This is not an inevitability, as attested to by distribution patterns and pricing formulas implemented in cities like Bogotá (Colombia), Phnom Penh (Cambodia), Porto Alegre (Brazil), and Santiago (Chile). Such pricing formulas entail means-tested prices or, preferably, nonlinear prices, with a very low ceiling for the first tranche of consumption. In Israel, where water is tightly monitored and generally scarce—despite running highly efficient desalination plants, recycling wastewater on a grand scale, and trampling Palestinians' rights to surface and groundwater—nonlinear pricing is the rule. Nevada also charges nonlinear prices in an effort to allocate as efficiently and fairly as possible the rather modest share of Colorado River water that the Colorado River Compact of 1922 granted to a state that, at the time, had fewer than 100,000 inhabitants. Necessity is the mother of wisdom.[3]

California has strong reasons to learn from these approaches. However, even after years of severe drought and with the looming prospect of ever-increasing water scarcity, most California communities seem unable to properly price water consumption (for exceptions, see the discussion in chapter 6 of Orange County's water revolution). As long as this economic infirmity remains uncorrected, there will be no sustainable solution to the state's water crisis. If publicly managed instruments, such as nonlinear prices, are not welcome in California, market-based instruments might be implemented instead, as they are in the Australian Murray-Darling Basin (see the Introduction to the book); that, however, would require an upheaval in the jumble of inefficient, illiquid, and inequitable water rights that plagues California.

All over developing countries, lack of sanitation in cities is an even more prevalent—albeit, related—problem. "The sewer is the conscience of the city," wrote Victor Hugo in *Les Misérables*. In his time, extended sanitation networks were being built in London, Paris, and then New York. It is believed that those networks contributed to about 50 percent of the fifteen- to twenty-year increase in life expectancy enjoyed by inhabitants of those cities between 1870 and 1910. As the aforementioned UNDP report (2006) stresses, "Water and sanitation are among the most powerful preventive medicines available to governments to reduce infection diseases" (6). However, investing in sanitation for the benefit of the poor rarely appeals to the ruling elites; thus, most public authorities in the developing world neglect sanitation. Note that there are a few significant exceptions—for example, in parts of Bangladesh, Brazil, and India—indicating that cost-effective realizations that are not unduly capital-intensive are feasible.

WASTE IS EVERYWHERE AND KNOWS NO IDEOLOGICAL FRONTIER

HOUSEHOLDS, INDUSTRY, AND AGRICULTURE

Water is often wasted as if it were a free good, which it less and less is. This issue weighs heavily on economic development and welfare, but it also opens encouraging prospects, as potential savings are proportional to quantities wasted. In addition, many of these savings are neither technically difficult nor costly to implement.

There is waste in cities, with leaks along the mains and at the taps, as well as indiscriminate use of first-quality potable water. Even more detrimental is the destruction of ecosystems that contribute to the storage and purification of water, as occurs around large Indian cities like Delhi and Bangalore, where lakes and swamps have been reclaimed, resulting in acute water-storage problems. In southern China, systematic destruction—or, at best, neglect—of lakes, swamps, and all sorts of pools and drains goes a long way toward explaining why the July 2016 floods were so devastating. In São Paulo, Brazil, the indiscriminate destruction of nearby forests and aquatic ecosystems dramatically magnified the effects of a lasting nationwide drought. In Isfahan, Iran, the once generous Zayanderud ("River of Life") has been reduced—except in times of flooding—to a dusty ribbon weaving through the city.

There is also waste from international trade. Indeed, biased prices for water (as well as for the electricity used to pump water) induce exports of goods with a high content of water (known as *embedded water*) from countries where water is scarce to countries where it is less so, possibly much less—for example, from China or Vietnam to northern Europe. It is thus essential to take into account embedded water in recipient country water consumption. In Britain, for example, the direct average daily consumption per capita is about 150 liters, which is low compared with most developed countries (for consumption in U.S. cities, see Box 2.1). However, the average daily consumption of embedded water in Britain is more than twenty times higher than that.

Industrialists in developed countries are increasingly concerned with the availability of water. Already, some power plants have had to shut down in periods of drought. These industries are generally efficient at saving, purifying, and recycling water, though the coal and shale hydrocarbon industries are notorious exceptions (see chapter 3), as is the food-processing industry. However, in emerging and developing countries, waste from industrial sources, mingling with household waste and agricultural runoff, can have horrendous effects, as exemplified by the situations in China and India.

In this respect, the Ganges River is emblematic. Of course, for millennia, it had been emblematic for a different reason—"eternally pure" is one of its sacred Sanskrit names. As Diana L. Eck, professor of comparative religion and Indian studies at Harvard University and author of *India: A Sacred Geography* (2013), put it, "A river seen as a source of salvation by the millions who include it in their daily rituals is now itself in need of saving. The issue is not just environmental; it is a cultural and theological crisis" (135). Indeed, from a bed supercharged in heavy metals and pesticide residues to surface waters nurturing all sorts of bacteria, many of which are resistant to antibiotics, the river is now toxic, carcinogenic, and home to all sorts of transmissible diseases. Conducting funeral rites in Banāras has become a highly hazardous endeavor. However, this condition is not completely irreversible, as has been shown in places as diverse as Chicago, London, and Manila. But saving the Ganges will require a revolution in the governance of India, just as, in a different context, a revolution in the governance of China is a precondition to significant improvements of Chinese rivers.

Compared with other uses of water, the demand from agriculture is by far the largest, using more than two-thirds of the total quantity of water consumed on earth. About half of these two-thirds is lost to leaks in irrigation networks, to evaporation, and to excessive quantities of water being brought to plants. As a consequence, some irrigated lands have become heavily salinized (Foley et al. 2005). An extreme example of wasteful irrigation is found in India. Well-off farmers pump water from the aquifer—water is free for them, and the electricity to run the pumps is heavily subsidized. Consequently, the aquifer is shrinking at a rapid pace of up to a meter per year in certain places. The traditional wells are now dry, and poor peasants are completely dependent on increasingly irregular monsoons. In the spring of 2016, after two poor monsoon seasons, water was severely lacking; for the majority of farmers, who have no access to ultra-deep pumping, the consequences are dramatic. The situation is extreme in the state of Maharashtra, west of Mumbai, where poor farmers have committed suicide by the hundreds, while companies protected by corrupt politicians in Delhi continue to pump huge quantities of water for their thirsty sugarcane crops. Even the Taj Mahal's foundations are in danger of collapsing as the aquifer shrinks.

Although depletion of essential aquifers is an acute problem in India, it is far from being exclusively an Indian problem. A victim of similarly deep-rooted mismanagement, as well as the effects of climate change on Himalayan glaciers and monsoon regimes (see chapter 4), is Pakistan. According to Khawaja M. Asif, Pakistan's minister for water and energy, "In the next six to seven years, Pakistan can be a water-starved country" (quoted in an interview by Salman Masood in the *New York Times*, February 12, 2015). Water availability per person per year is already at 1,000 m^3, which is a "water-stressed" level by international standards, down from 5,000 m^3 in 1947, the year of independence. India is not yet down to this level, but at 1,400 m^3, its situation is only slightly better, and it is worsening.

In Iran, water is being wasted by farmers irrigating fields without restraint and by engineering firms, owned by the Islamic Revolutionary Guard Corps, damming most rivers (or at least those that are not yet completely dry). The fate of Lake Urmia, the Iranian counterpart to Utah's Great Salt Lake, is reminiscent of the destruction of the Aral Sea during the communist-era rush to cotton in Central Asia, with similar dust storms blowing in from the parched lands. Most of Iran is now in the grip of what looks like a never-ending drought. Water-management routines, which essentially boil down to depleting aquifers, don't prepare the country to face a desertification trend that will get all the more severe as climate change advances. Desertification is also on the march in Kerman province, in southeastern Iran—an area that is sometimes referred to as "pistachio land." Farmers there waste subsidized water by lavishly flooding their trees in the middle of the day. Some years ago, when one farmer invested in a drip irrigation system, he became the laughingstock of his neighbors. Yet, his fields are now the sole green spot in a landscape of dead trees, as he is the only one who can make do with today's meager water allocation.

Wasting water in agriculture (and in some other sectors) is by no means confined to developing countries (see, for example, Box 2.1). Just as with other key natural resources, wasting water knows no ideological frontier.

BOX 2.1 TEXAS: WASTE AT WILL?

The perverse relationship between Texans and water is illustrated in both towns and cities.

High Plains farmers, who depend on extracting water from the Ogallala Aquifer, should be mindful of the quantities they pump, especially as they sit on the fringe of the aquifer, where there is much less water left than in the central part of the aquifer (e.g., in Nebraska). However, these farmers don't easily accept any kind of discipline of their consumption. According to Texas law, these farmers privately own the portion of the aquifer beneath their land (whatever that might mean); they interpret such ownership as a right to pump at will. Accordingly, they have decided to fight in court what they see as a property rights violation.

In 2011, the High Plains Underground Water Conservation District introduced a regulation requiring that new wells be equipped with meters and that withdrawals from these wells be capped. Reacting to the farmers' outcry, which was vocal despite the exclusion of existing wells from the regulation, the district board, which consists of five locally elected officials, voted in February 2012 to accept a moratorium. Entrenched interests, flawed views, bad habits, and weak institutions induce in a highly developed region the same logic of waste as is seen in northern India and Pakistan, a logic that is at work in many parts of the world.

(*continued*)

It is all the more striking that efficient ways of using water for irrigation are demonstrated not far from the High Plains in the Texas Panhandle. There, under the guidance of the Panhandle Groundwater Conservation District, which succeeded in imposing limits on pumping from the Ogallala, several farmers have changed their ways, roughly halving the amount of water needed to grow corn, the most common crop in the district. These farmers delay planting to take advantage of the rain that usually falls at the end of the season, they give each plant more space, and they leave old stalks in the fields to keep more moisture in the soil. Some of them also adopt remote-sensing technologies pioneered by farmers in Nebraska (see chapter 6).

Ignoring the natural constraints of their semiarid (or even arid) environment, cities in Texas and some neighboring states use water as if they had resources matching those available in Washington state or British Columbia. Profligate use of water is even made compulsory in well-off Texas suburbs, with bylaws enforced by neighborhood associations requiring immaculately green turf lawns and automatic sprinkler systems. Saint Augustine grass is particularly popular there, as it makes lush, deep green lawns. But this grass is native to damp, tropical ecosystems, which means it requires regular, generous watering in places like Dallas and Phoenix.

In Dallas, Texas A&M University maintains a demonstration garden where visitors discover with amazement that lush palisades of *Zoysia* grass need watering during the summer only once a week. This garden attracts many visitors, albeit fewer imitators in Texas than in Nevada, where a more realistic water policy is being implemented. It is no wonder that the authors of a study encompassing seventeen cities, commissioned by the Texas Water Development Board, reported: "Outdoor water use as a percentage of total use figures ranged from a low of 20 percent in Houston to a high of 53 percent in Tyler, with a weighted average of 31 percent across the State" (Hermitte and Mace 2012, 22). This number for Dallas is 40 percent, and the average daily consumption per capita (indoor and outdoor) there amounted to 110 gallons in 2013. In the affluent Dallas suburbs of Highland and University Parks, Sunnyside, and Southlake, the average daily consumption per capita is more than three times higher, and the outdoor water use increases exponentially with the wealth of the residents.

Although a comparison with Boston (which had an average per capita consumption of 40 gallons per day in 2013, all uses included) would not be appropriate due to the different meteorological conditions, a comparison with Santa Fe is appropriate. In 2013, the average daily consumption per capita in Santa Fe was 65 gallons, including 25 gallons for outdoor use; the price for high-volume users is almost four times what it is in Dallas and five times what it is in Phoenix.[4] (Prices in Santa Fe increase significantly with amounts consumed, as the tariff is nonlinear.) Thus, economic incentives do work, albeit less so with wealthy people.

PHARAONIC PROJECTS ARE NO SUBSTITUTES FOR GOOD GOVERNANCE

Rather than reorienting an unsustainable mode of development and saving water that is wasted in huge proportions, Chinese authorities carry on pharaonic projects. For example, to remedy the lack of water in the north, they divert water from the south. This policy is based on Mao's vision, inspired by the role of the Beijing-Hangzhou Grand Canal in Chinese history, from the Sui and Tang dynasties in the sixth to ninth centuries to the Ming dynasty in the fourteenth to seventeenth centuries. Mao's vision is now materializing in the South-North Water Transfer Project, which is meant to divert to the north water from the Yangtze River basin. Central planners decide on the amounts of water to be diverted, based on water flows in the Han River (the main tributary of the Yangtze) that were measured over a period extending from the mid-1950s to the early 1990s. Since then, water flows have dropped, partly because of unexpectedly severe droughts and partly because of increasing local withdrawals. All over the Yangtze basin, pollution has shot up. Nevertheless, central planners have not adjusted the scale of the project. Will the main result of the South-North Water Transfer Project be a Yangtze basin as degraded as the Yellow River basin, which the project was originally supposed to help revive? For now, the water that has already been diverted has gone mainly to large cities in the northeast—particularly, Beijing—at huge ecological costs to the Yangtze and Han rivers, not to mention the huge human costs that have occurred along the waterways built for carrying the diverted water. In this process, more than half a million people have been displaced, most of them peasants pushed to lower-grade farmland.

With 20 percent of the world's population and a growing economy, but only about 6 percent of the world's freshwater resources, it is no surprise that China faces serious problems of water availability and quality, especially as its water resources are unevenly distributed in space and time. But there is no point in trying to find solutions in grandiose schemes that spread problems rather than water. Instead, efficiently dealing with China's daunting water problems requires reforming a mode of development that is systematically wasteful and unsustainable. In Qiu Xiaolong's novel *Don't Cry, Tai Lake*, Chief Inspector Chen Cao clearly grasps the essence of China's water crisis:

> Look at them. Paper mills, dyeing factories, chemical companies, and what not. In the last twenty years or so, those plants have sprung up like bamboo shoots after the rain. Now they make up more than 40 percent of the city's total economic output. Relocating them is out of the question, there are too many of them. The local officials aren't eager to do anything about it. . . . Sure, there's a city environmental office, but it exists only for appearance sake. Some of the factories are equipped with wastewater processing facilities, but they generally choose not to operate those facilities. The cost of doing so would wipe out their profits. So they have the facilities for the sake of appearances, but continue to dump waste into the lake in spite of the worsening crisis. (2012, 29)

One might think that the overall restructuring of the Chinese economy toward less capital-intensive activities would eliminate most of the outdated polluting factories. Many of them, however, hang on as "zombie factories," which no longer create real value even as they continue to deplete and pollute water, soil, and air. There are ways to overcome the water crisis in China, but implementing them requires deep changes in the system of political and economic incentives; it requires a mode of development that is sustainable, participatory, and equitable, not a confiscation of power and resources.

Water is both a paramount problem in itself and evidence of social weaknesses and perversities. In China, as in many parts of the world, the potential for water savings is huge. The appropriate technical and managerial instruments to realize this potential exist and have been tested in various contexts (see chapter 6). However, appropriate political and economic institutional arrangements are prerequisites for a broad use of these instruments, and time is running out as climate change is around the corner, already disrupting water regimes by eating at Himalayan glaciers.

WATER IS FOR FIGHTING OVER

Fights over water rights in the southwestern United States are proverbial and ongoing. As just two examples, a dispute between New Mexico and Texas has reached the Supreme Court; Colorado, Kansas, and Nebraska are at a more preliminary stage in a case pitting Kansas against the other two, which are accused of pumping too much water from the Republican River. Less known are the increasingly bitter fights in the southeastern United States, a region where water availability used to be the envy of Westerners. Even there, droughts have become more of a threat due to the population explosions in Atlanta, Georgia, and along the coasts of Florida; even Alabama is developing. More inhabitants, increasing incomes, and economic growth generate increasing water demand at a time when nature has become less generous. Alabama, Florida, and Georgia are now in permanent judicial conflict about the allocation of the insufficient waters from the Chattahoochee and Tallapoosa River basins. However bitter the conflict may be, ultimately it will be arbitrated in court or by Congress.

When it comes to international conflicts over water, arbitrage is rather more problematic. Egypt, where it almost never rains, desperately clings to its large share of the Nile waters (slightly more than 70 percent, according to the 1959 Nile Waters Agreement with Sudan), despite increasing pressure from countries upstream, which require more water for their economic development and rapidly increasing populations (Egypt: 81 million, Sudan: 44 million, Ethiopia: 83 million, for a total of 208 million in the region; region totals in 2025 and 2050 are expected to reach 272 and 360 million, respectively). Egypt has the disadvantage of lying downstream but the advantage of a desperate situation that makes credible (see Schelling 1981) a threat of resorting to military force, as its army and air force are dominant in the region. However, countries from outside the region—China, India, Korea, and Saudi Arabia (not to mention some U.S. pension funds and private universities)—are now joining the game. These factions are acquiring large tracts of agricultural land all along the Nile, where they intend to practice "modern" forms of agriculture that require a great deal of water to materialize the expected yields. They seem to consider that they have acquired the rights to the water resources along with the rights to the land.

Is there a solution to this that does not involve violent confrontation? The Declaration of Principles signed in March 2015 by Egypt, Ethiopia, and Sudan looks like an overture toward peacefully protecting the interests of the downstream countries (i.e., Egypt and Sudan) once the giant Grand Ethiopian Renaissance Dam on the Blue Nile is completed. There will, however, be many opportunities for these nations to disagree when it comes to the actual joint management of the entire basin.

Vietnam is increasingly unable to cope with the effects of transformations upstream on the Mekong. The main effects are the destruction of essential fisheries, the deficit in nutrients, and the disruption of flood regimes that are crucial for farming in the Nine Dragons delta, where eighteen million Vietnamese live. The effects of the dams built, without any mutual consultation, on the Chinese upper reaches of the river are bad enough; the materialization of the Xayaburi Dam in Laos (the project is Thai, and most of the electricity would go to Thailand) on the river's middle course will dramatically disrupt the water regimes, the circulation of sediments, and the ecological equilibrium. It should rationally be a casus belli for Vietnam.

Syria and Turkey are increasingly depriving Iraq of water from the Euphrates and the Tigris. The effects of this water deficit are compounded by an irreversible advance of salinization, which started in the Shatt al-Arab delta and is progressing upstream. Although Iraq is in no position to consider military options, it might consider swapping oil for water. However, there is a real danger that nothing productive will materialize before drought, salt, political chaos, and surrounding wars have turned the country into a desert.

The extent to which the main aquifers in the world are being depleted is now known with some degree of certainty. Gravity measures made by a pair of satellites, whose respective speeds vary with gravitational variations beneath them, allow scientists to record water levels in aquifers. The most recent results, presented and discussed in Richey et al. (2015), have led to the conclusion that between

1.5 and 2.0 billion people live in areas where groundwater resources are under serious threats. Aquifers under major threat include California's Central Valley Aquifer; the aforementioned Ogallala Aquifer at least in Texas; the North-Western Sahara Aquifer System, shared by Algeria, Tunisia, and Libya; the Arabian aquifer, mainly in Saudi Arabia; the Indus and Upper Ganges aquifers, shared by India and Pakistan; and the Yellow River Basin aquifer in northern China. Most aquifers (both national and international) suffer from the free-rider effect. Some are depleted by misguided public policies, such as subsidies in India for indiscriminate pumping from the aquifers or the Saudi government's ambition to grow cereals on a large scale (see discussion later in this chapter).

It is well known (and not merely theoretical) that appropriate forms of cooperation can overcome free-rider behaviors. Relevant provisions in the Guarani Aquifer Agreement between Argentina, Brazil, Paraguay, and Uruguay have proved it can be an effective deterrent. Of course, it should be noted that this agreement has been much facilitated by the fact that the Guarani Aquifer is not under immediate threat of depletion or pollution; thus, for now, the positions are not rigidified by entrenched interests.

SOIL AS AN ECOSYSTEM

COMPONENTS AND INTERACTIONS

In his masterful handbook *Soil in the Environment* (2008), Daniel Hillel explained how diverse a system is a soil, what attributes make a soil fertile, and to which threats it is exposed:

> The capacity of a soil to serve as a favorable medium for plant growth depends on several interrelated attributes. The soil must be porous and permeable enough to permit the free entry, retention, and transmission of water and air. It must also contain a supply of nutrients in forms that are available to plants but that do not leach too rapidly. The soil must be deep and loose enough to allow the roots to penetrate and proliferate. In addition, the soil must have

> an optimal range of temperature and acidity, and be free of excess salts or of toxic factors. . . . It is home to a varied and interdependent biotic community consisting of myriads of organisms, including microscopic plants and animals, all engaged in a complex set of complementary functions. . . . It is the very diversity of life that imparts to the soil the ability to respond to and recover from perturbations such as episodes of drought and flooding, as well as contamination by pollutants; that is to say, biodiversity enhances the soil's stability and resilience.[5] (151)

Hillel described demanding conditions for a soil to be healthy. Even when these demands are all met, it is all too easy to undermine the balance, with dire consequences for the soil itself and for the crops that the soil is supposed to sustain. Whether from natural effects or human actions, the balance is indeed too often undermined in both the short and long term—a tragedy at the heart of agriculture. To see why and how this tragedy occurs, we must have a minimal understanding of how soil functions.

Let's first consider nutrients. Specialists differentiate between macronutrients and micronutrients. The main two macronutrients are nitrogen and phosphorus (calcium and potassium also play significant roles). Nitrogen is an essential component of the nucleic acids and proteins in plants, as well as of the chlorophyll, without which photosynthesis is not possible. Phosphorus is also an essential component of nucleic acids, as well as of cell membranes and the molecule that supplies the energy that sustains metabolism. Micronutrients such as iron, manganese, copper, and chlorine are needed as trace elements; if a plant suffers deficiencies in micronutrients, its ability to use the macronutrients may be severely curtailed.

There is a remarkable contrast between carbon dioxide and nitrogen in their undertakings with plants. Carbon dioxide, which makes up less than 0.05 percent of the atmosphere, is easily captured by plants during photosynthesis. Nitrogen, despite constituting more than 70 percent of the atmosphere, is not captured directly by plants, and not at all by most of them. Leguminous plants (legumes), such as peas, beans, and soybeans, conspire with specific bacteria growing on their roots to succeed in capturing atmospheric nitrogen.

(This process of symbiosis may also be artificially initiated by inoculating seeds with the required bacteria, as is done in Brazil's soybeans to save on chemical fertilizers.) There are also "fertilizer trees," like the acacia, alder, or tropical desmodium and casuarina.[6] These "nitrogen fixers" benefit from the amounts of nitrogen they capture while also enriching the soil around them with nitrogen to the benefit of other plants that either coexist with them or are grown in rotation on the same tract of land. Another way to make nitrogen available in the soil is to return organic matter to the soil in the form of plant residues and animal manure, which are transformed by microbial decay. Yet, it is the very purpose of agriculture to take away plant and animal content; hence, only a small fraction of the nitrogen that has been used is brought back to the soil. In the absence of leguminous plants and fertilizer trees, the deficit in the soil can be made up by the application of nitrogen fertilizers produced by the chemical industry. It is manifest that such fertilizers are very efficient in boosting crop yields; as they are systematically overused in developed and emergent countries, they lead to serious air, soil, and water pollution (see discussion later in this section).

Apart from recycling phosphorus contained in plant residues and animal manure, the only widely available way to provide plants with phosphorus is through the application of superphosphates, which are mined in a small number of countries. Such dependence on a rare substance, not to mention the depletion of the mines themselves, generates problems that are specific to phosphorus availability. (To a certain extent, potassium is mired in similar problems.)

Along with nutrients, bacteria and microfungi also play essential roles within the soil ecosystem. They are present naturally in stunning numbers (1,500 kg/ha of bacteria is not uncommon). The main contribution of bacteria is to decompose organic matter and recycle nutrients. Certain bacteria can play specific beneficial roles, as they do with leguminous plants, but other bacteria are detrimental to plants (see chapter 1). Likewise, there is a great variety of microfungi, not all of which are beneficial (see chapter 1). Some have unexpected characteristics—for example, long, narrow microfungi are able to penetrate regions in the soil that are inaccessible to plant roots, allowing plants to absorb extra water and nutrients from the soil.

Whereas the roles of bacteria and fungi are not always appreciated, earthworms are often hailed as contributors to the maintenance and improvement of soil quality. They tend to proliferate in good-quality soil (a few million per hectare is not uncommon), sustaining beneficial feedback loops for plants. Earthworms ingest organic debris, like leaves and dead roots, and produce nutrient-rich casts of which plants can make use. They do so while tunneling through the soil, hence forming networks of pores that improve air and water circulation. Their reputation, to which Aristotle contributed when he called them "the intestines of the earth," clearly is not usurped.

POLLUTANTS AND EROSION

Organic and inorganic pollutants from all origins are aggressing soil and other natural environments in ever-increasing numbers and quantities, emphasizing even more the importance of bioremediation. As an example, the actions of living organisms, such as specific plants, have the remarkable ability to absorb contaminants (such as dangerous heavy metals) or to degrade them into less harmful substances.

Soil is also under threat of water and wind erosion. Erosion is a natural process that human actions can either accelerate or contain. Water erosion results from the impact of striking raindrops and from the pressure of water flowing over the soil. As for the effect of wind, the more strongly it blows and the more desiccated the soil (under drought conditions, in particular), the greater the volume of soil lost.

While on-site impacts of erosion (where erosion takes place) are overwhelmingly negative, off-site impacts can be either negative or positive. They are negative when pollutants are disseminated through erosion into rivers or in reservoirs that get clogged by sediments and contaminated by pollutants. However, an example of positive impacts can be seen in Egypt, where, for millennia, farmers depended on the nutrient-rich sediments brought by the Nile River to sustain the fertility of their fields. Less well-known, but no less massive, is the aerial transport of fertile dust across the Atlantic Ocean to the Amazon basin from the bed of the long dried-up Lake Mega-Chad in Africa, which was once as large as all of the Great Lakes of North America combined (Bristow, Hudson-Edwards, and Chappell 2010).

To minimize water erosion, the soil should be neither too compact, so that more water can infiltrate, nor too destructured (e.g., as a result of deep tillage), so that the topsoil is not too easily taken away by water flow. Retaining crop residues as mulch on the soil improves soil structure, keeps moisture on the surface, and protects against the mechanical effects of water and wind. In Latin America, zero tillage is adopted on a large scale (slightly more than 40 percent of the arable land), and mulch coverage is systematic. Another efficient way to guard against the worst effects of erosion is to mingle perennial plants and trees with the crop on the field (see chapter 8). This tactic was not followed on the American prairies, which were converted to cereal fields during the 1920s, with the Dust Bowl ensuing soon after in the 1930s.

Erosion manifestations, with desertification as an extreme form, are nothing new; erosion was a curse of ancient Greece, as just one example. Today, however, they take global proportions, from Mongolia and northern China to the countries bordering the Sahara to the western United States. Climate change will not help, as illustrated in Romm (2011) and on the UN website "Desertification, Land Degradation, and Drought."

PROGRESS TOWARD A DEAD END

LACK AND EXCESS OF NUTRIENTS

Lack of nutrients in African soils induces a spiral of soil impoverishment and degradation, dramatically reducing yields and ending in sterility. A hectare of cultivated soil in sub-Saharan Africa gets, on average, about 10 kg of nitrogen fertilizer a year, and even less in many places (Gilbert 2012). As it is the main nutrient, nitrogen is a good indicator of scarcity. However, bringing more of it to the soil, with no regard to the balance of all the necessary nutrients, does not help, even if this is too often what happens. At some point, desperate farmers look for fresh soils to reclaim from forests or from other valuable ecosystems; in doing so, they create other forms of ecological imbalances.

At the other extreme of fertilizer application are the industrialized and emergent countries—China, in particular. In Europe and the United States, farmers apply an average of 200–250 kg/ha of nitrogen fertilizers per year, of which only 20–30 percent is taken up by plants (Sutton and Bleeker 2013). In China, this number is as high as 400 kg/ha, with even poorer intakes (Zhang et al. 2013; Liu et al. 2013). In fundamental respects, Chinese agriculture magnifies the imbalances in which industrial agriculture in America and Europe is trapped.

The benefits of nitrogen fertilizer application are indeed strongly nonlinear (this is also true for the other nutrients—especially for phosphorus). Thresholds are relatively quickly reached, beyond which intakes plummet. Farmers are nonetheless encouraged by salespeople working for fertilizer firms, as well as by experts from agriculture administrations and specialized research and development (R&D) bodies, to think otherwise. In China, for example,

> before the 1990s, scientists, government and extension staff encouraged farmers to increase synthetic fertilizer inputs to increase yields and feed an increasing population. Persuading farmers to limit fertilizer inputs is now difficult because many of them still hold to now traditional opinions that higher crop yield will be obtained with more fertilizer. (Ju et al. 2009)

As fertilizers are heavily subsidized and their health and environmental detrimental effects are not penalized as they should be, farmers are not encouraged to change their views. Economic incentives are completely upside down.

Quantities in excess of intakes are not just lost; they severely pollute water and air and are the source of potent greenhouse gases. Excessive amounts of Nr (i.e., the reactive compounds made from nitrogen at great energy expense)—in particular ammonium—find their way to water bodies, saturating rivers, lakes, and aquifers with nitrates, making the water dangerous for human consumption. Although these compounds are meant to be nutrients for crops, they also feed algal blooms, causing freshwater and coastal dead zones by hypoxia (disappearance of oxygen that is consumed as the algae decay), where all plant and animal life becomes impossible (see chapter 1). Excess amounts of nitrates remaining in the soil also act as pollutants; in particular, their persistence sustains a trend of acidification (Guo et al. 2010) and poisoning of the soil.[7]

COLLATERAL DAMAGES OF BIOCIDES

In developed and emergent countries, chemical biocides (pesticides, herbicides, fungicides) are used with no less recklessness than are chemical fertilizers. Their side effects are even worse than those of fertilizers, as they devastate biodiversity in the soil and the water bodies into which they leach (Beketov et al. 2013). They are extremely dangerous for farmers' health and, more generally, for public health (see chapter 7). For instance, pesticides can reach babies in utero, and residues are ubiquitous in food processed from industrial agriculture.[8] Biocides are detrimental to all sorts of animals, birds, pollinators (see chapter 1), and even earthworms (Hopwood et al. 2013).

Biocides often prove more efficient at inflicting collateral damages than at hitting their targets. The fact that they often have negative effects on the populations of the natural predators of these very targets doesn't help either. It is no surprise that the chemical industry has been led to promote its biocides even more aggressively and dishonestly than its fertilizers (see the atrazine case in chapter 7). Despite their capacity at dealing with emergencies, biocides, when used routinely, tend to cause their own demise by prompting a Darwinian selection of resistant pest strains (see chapter 6; Palumbi 2001). In the ensuing arms race, even though the producers' profits might be enhanced, the health of farmers, the public, and the environment all suffer.

Yet, chemistry might take another turn, as the Royal Society Report (2009) hints: "There is potential for a novel class of crop protection chemicals that are fundamentally different from those most widely used at present. The novel compounds would resemble chemicals present in plants that activate or prime natural resistance mechanisms and because they do not target pests and pathogens directly they could have environmental advantages over currently used compounds" (30). The report also advocates the broader potential of the integrated management of pests and diseases based on biological interactions (see chapters 6 and 8).

LAND GRABS: A ROAD TO UNSUSTAINABILITY

To become less dependent on food imports, Saudi Arabia began to grow cereals on its own territory, despite the paucity of water. But the water requirements proved so large with respect to the country's limited resources that, after a few years, outsourcing was substituted to home growing. Various public bodies and private companies started buying large tracts of arable land in Africa or taking long leases. Other countries, such as China, India, Korea, and Malaysia, made similar moves in Africa and in poor Asian countries like Cambodia, Laos, and Papua New Guinea. The main U.S. and European traders in agricultural commodities also started buying this land—for example, Louis Dreyfus in former French colonies. Some U.S. pension funds and rich universities joined the fray in Africa and then set their sights on South America.

In general, national and local authorities welcomed these moves for various reasons—certainly not all honorable—and saw to it that the newcomers could take hold of their land to their satisfaction. When, as is frequently the case, the land is cultivated by farmers with mere customary tenure rights, with the land being formal property of the state, the problem has been solved with massive expulsions. Although some of the expelled farmers have been hired by the new owners, most of them either have been regrouped in "new villages" on poor farming land or have joined the slums of nearby cities.

From 1995 to 2016, the government of Ethiopia leased or sold about 7.0 million ha to foreign investors; in turn, more than 1.5 million smallholders were moved off their land. In a particularly dramatic turn in the southern Gambela region, about 40 percent of cultivated land has been transferred to foreign firms. The Anuak indigenous community had been living on fertile soil in that region, from which they were uprooted much more by intimidation than persuasion, including arbitrary arrest, rape, torture, and murder as practiced by the Ethiopian police and military. Okello A. Ochalla, who was governor of the region when massacres were perpetrated, tried to oppose the mass violations of human and land rights in 2003. He then fled Ethiopia and became a Norwegian citizen. All the same, he was kidnapped on behalf of the Ethiopian government, imprisoned, tortured, and charged with terrorist activities.

The Cambodian government, which is in the orbit of China, is no less harsh in the way it clears land for new Chinese owners and for Cambodian cronies, regardless of property rights. In Laos, more land is now owned by foreigners (Chinese, Vietnamese, Thais) than is used for growing rice. Both people and ecosystems suffer.[9]

This is not the only form of unsustainability associated with land grabs. The new owners are, in general, interested in quick, high yields, which they expect to achieve through agricultural practices involving massive applications of chemical implements and generous irrigation, which is rather inimical to the original biodiversity and sustainability of the soil. Competition for water is set to raise particularly thorny problems, especially when the water bodies involved (rivers, lakes, aquifers) are shared among different countries. It is clear, for example, that the Saudis, the Chinese, and others in Ethiopia expect that enough water will come along with the land they own. Insofar as that water significantly reduces the flows of the Nile entering Egypt, severe conflicts are inevitable (see earlier in this chapter).

Finally, what will happen when not enough food is available in a host country? Will the inhabitants try to block exports of products from the farms owned by foreigners? Will they succeed? If so, will the foreigners discover that their property rights may not be exercised in all circumstances? Or will the national and local authorities see that these rights are exercised in all circumstances, whatever the human cost? In any case, we can forget about sustainability.[10]

3

ENERGY: AS LITTLE AS POSSIBLE

THERE ARE two fundamental concerns regarding energy production and consumption. First, how do we make available to people in developing countries—where energy availability is insufficient, inconvenient, and often dangerous—adequate energy supplies to efficiently meet their needs?[1] Second, how do we minimize, in all possible ways, fossil fuel consumption in developed and emerging economies? This second concern can be addressed by systematically disseminating the clean technologies most appropriate to the various circumstances of diverse users (see chapter 8) and by changing behaviors through appropriate incentives and regulations (see chapter 9).

It is generally agreed that, to avoid major climate disturbances, the increase in the earth's mean temperature should be kept to less than 1.5°C–2.0°C above the preindustrial level (see chapter 4). Thus, a relevant question is what should be done, as far as constraining greenhouse gas emissions is concerned, to actually remain under that upper bound? The prevalent answer is that cumulative carbon dioxide (CO_2) emissions should henceforth not exceed a total budget of 600–700 gigatons (GT) in order to have a 0.75 probability of not exceeding the limit (Meinshausen et al. 2009).[2] Under that constraint, it would be impossible to burn all proven fossil fuel reserves which are rather conservatively defined and which include neither nonconventional shale gas and oil nor Arctic reserves–as doing so would produce a total amount of CO_2 emissions somewhere between 2,500 and 3,000 GT. As pointed out by Nicholas Stern in the *Financial Times* (2011), "There is . . . a profound contradiction between declared public policy [i.e., less than +2°C] and the valuations of these listed [energy] companies, based on their fossil fuel reserves," which appear to assume that the world will not get anywhere near its targets for managing climate change.[3]

It would be a complete delusion to expect that physical constraints on the availability of fossil energy resources might help bring sobriety in time. There is no such thing in sight as peak coal, peak gas, or peak oil. As far as coal is concerned, it is well known that relatively easily exploitable reserves are available that might last for several centuries. In addition, it has recently become apparent that gas and oil are still available in far larger quantities than previously suspected, due to new discoveries and, even more so, to new technologies making it possible to pump from deposits hitherto inaccessible. These findings are "game-changing," said Khalid al-Falih, then CEO of oil company Saudi Aramco. "By pushing concerns about security of supply in the background, the shale boom helps encourage technology and policy choices that will lock in demand for oil. . . . We look at ourselves as being in this business for generations to come" (quoted in Crooks 2013). Yet, this is not good news for generations to come.

There is a pressing need on a global scale to produce and use energy more efficiently, rebalancing the mix away from fossil fuels. Unfortunately, the pool of available fossil fuels has been expanding at a pace not seen for decades, and low prices are attractive to consumers. This expansion is largely under the control of extremely powerful corporations, both private and public, and it has ramifications in all sectors of the economy, politics, and the media. According to the International Energy Agency (2015a), these corporations have spent far more on extracting fossil fuels, as well as on searching for resources and more efficient techniques to recover them, than what has globally been spent on energy efficiency and energy savings. And, of course, it is their intention to sell the stuff. Here is a contradiction

Table 3.1 World primary energy demand by fuel and scenario (in million tons of oil equivalent)

	2000	**2013**	**2040 (Current Policies)**	**2040 (New Policies)**	**2040 (450 Scenario)**
Coal	2,343	3,929	5,618	4,414	2,495
Oil	3,669	4,219	5,348	4,735	3,351
Gas	2,067	2,901	4,610	4,239	3,335
Nuclear	676	646	1,036	1,201	1,627
Hydro	225	326	507	531	588
Bioenergy	1,023	1,376	1,830	1,878	2,331
Other renewables	60	161	693	937	1,470
Total	**10,063**	**13,559**	**19,643**	17,934	**15,197**
Fossil fuel share	80%	81%	79%	75%	60%
Non-OECD share	46%	60%	70%	70%	69%
CO_2 emissions (GT)	23.2	31.6	44.1	36.7	18.8

Note: GT = gigatons; OECD = Organisation for Economic Co-operation and Development.
Source: International Energy Agency 2015b.

that is no less striking than the contradiction exhibited by Nicholas Stern, and it might spell the end of properly organized life on earth (see chapter 4)—an end that would result from too much of what almost everybody considers a good thing (see table 3.1).

Because environmental and climate threats will not just evaporate, priorities must be defined and implemented. Before considering them in later chapters, we present in this chapter a review of the main sources of energy—fossil, renewable, and nuclear.

Under the "current policies" and "new policies" scenarios, the climate blows up. Scenario 450 corresponds to the that laid out in Bloomberg New Energy Outlook 2016, which comes with a warning, introduced as the eighth "eye-catching finding" from the report: "On top of the forecasted $9.2tn investment in zero-carbon power, an extra $5.3tn is needed by 2040 to prevent power sector emissions rising above the IPCC's 'safe' limit of 450 parts per million [i.e., to remain under the +2°C limit]."

How do we avoid that? How do we significantly improve upon this scenario in terms of CO_2 emissions? We do so by (1) making it more difficult (through regulations) and more costly (through carbon prices) to burn fossil fuels; (2) maximizing the efforts in energy efficiency and savings; (3) maximizing the efforts of disseminating efficient clean technologies, especially throughout the developing world; and (4) capturing CO_2 not only in power plants but also from ambient air (on these points, see chapters 6, 8, and 9). At least, this is what we should do.

FOSSIL FUELS

COAL

Coal is the dirtiest fossil fuel—matched only by bitumen from tar sands—both as a local and regional pollutant and as a CO_2 emitter. Beijing, Delhi, and other Asian cities are reproducing the extremes of smog that were the plight of London during the first half of the twentieth century. Even for residents of the United States, the negative health effects produced by the coal industry are estimated at more than $30 billion annually (Buonocore et al. 2016). In the Balkans, a region of Europe heavily dependent on coal, the devastations evoke those seen in the coal-producing provinces of China. The co-benefits of reducing CO_2 emissions by scaling down the production and consumption of coal are significant indeed.

Coal produces more CO_2 emissions than gas and oil (provided that oil has not been extracted from tar sands and that methane leaks from gas extraction, transport, and consumption are minimized). In fact, in the United States, coal consumption is declining at a pace that would have been unthinkable at the beginning of the century, though this is mainly due to the substitution of shale gas. Stricter regulatory constraints on coal-fired power plants have also contributed to coal decline in the United States.

Some countries in the EU have scaled up the production of electricity from coal-fired power plants in order to compensate for the phasing out of nuclear plants and to make up for the intermittency of the production from a fast-increasing number of wind farms and solar panels. (Being less polluting and easier to turn on and off, gas-fired power stations are preferable, but in Europe, gas is relatively expensive.) Germany is on the forefront of this trend back to coal. In 2015, coal contributed 45 percent of total electricity produced in Germany, whereas renewable sources contributed 35 percent. Although the development of renewables since the beginning of the century has been remarkable in Germany, the growth in the amount of electricity produced from coal has also been sustained, with new coal-fired power plants connected to the grid each year.

Germany has vast reserves of lignite (also called *brown coal*), which is the lowest grade of coal being mined. The communist government in what was then East Germany depended heavily on lignite for both domestic and industrial uses; the resulting air, water, and soil pollution was horrendous. When Germany was reunified in 1990, most inefficient lignite-fired power plants were closed, and the use of lignite fell dramatically. But now lignite is back. In 2015, it contributed 25 percent of total electricity produced in the whole of Germany, making it the single most important source of electricity production in Germany. CO_2 emissions from lignite are particularly high per kilowatt-hour produced, and the contribution of lignite-fired power stations to the recent increase in Germany's CO_2 emissions is thus outsized.

As mentioned earlier, natural gas would do a significantly better job as a complement to renewables. In Europe, however, it cannot compete in prices with lignite and U.S. coal, the latter of which is sold at less than $100 per ton within Europe. The opposite is true in the United States, where coal can no longer compete with gas. U.S. coal producers freely sell their product abroad, but U.S. gas producers only received the first export licenses as late as the end of 2015.

With the technological progress in electricity production and energy storage (see chapter 6) from wind and solar, as well as social and political pressures, the share of coal is set to decrease. Anticipating that shift, E.ON SE, the largest German energy utility, has decided to spin off its coal-based activities, concentrating its resources on the development of renewables, networks, and services. In December 2015, RWE, the second largest energy utility, followed suit.

These developments, however significant, pale in comparison with China's contribution to the twenty-first-century coal renaissance. During the first decade of this century, China's total coal demand was multiplied by 3, reaching 3.5 billion tons in 2010. (During the same period, world demand increased from 5.0 to 8.0 billion tons.) In China, the 4.0-billion-ton ceiling was reached in 2012, according to statistics officially revised upward in October 2012. Since then, demand for coal has been flat or modestly decreasing. Electricity production requires huge quantities of coal, with about 80 percent of the electricity consumed in China produced in coal-fired power plants, many of which have poor thermodynamic conversion factors. Graft has been keeping pace with the growth of the industry, to the point of triggering, in 2014, a wave of investigations by the Central Commission for Discipline Inspection into alleged high-profile corruption cases linked to coal production and trade within the Chinese Communist Party.

According to projections by the International Energy Agency (IEA 2016), by 2040, more than two-thirds of China's electricity production will be from coal, despite dramatic investments in renewable sources unparalleled anywhere else in the world. If these projections materialize, so will a severe climate boomerang effect. However, if the Chinese economy's rate of growth continues to decrease, as it did in 2015 and 2016, and if investments in renewables are even larger than anticipated and don't remain idle for being refused access to grids (too often, coal-fired power plants enjoy preferential access to the grid), coal consumption and CO_2 emissions might be significantly reduced. There are thus serious contradictions and uncertainties when looking to the future.

China's coal demand is met mostly by domestic mines, even as imports from Australia, Indonesia, and a few other countries have been quickly increasing. Coal now provides more primary energy in China than does worldwide all the oil from the Middle East. Although far behind what is seen in China, India's coal demand has increased by about 80 percent during the past decade, illustrating a trend in several Asian countries that prompted World Bank president, Jim Yong Kim, to formulate a stern warning: "If all the new coal plants on the books earlier this year were constructed—especially in Asia—it would be impossible to stay under two degrees."[4]

NATURAL GAS

When properly handled, natural gases are significantly less polluting than coal. This results from their respective chemical compositions and from the greater efficiency of gas-fired power plants. Still, combustion of natural gas is not emission-free; it emits about half the amount of CO_2 emitted by coal for the same amount of energy produced. Moreover, natural gas is often not properly handled, which is all the more detrimental because most of the commercial natural gas is methane (CH_4), which is vastly more potent than CO_2 as a greenhouse gas (see chapter 4). If methane leaks are too large, they annihilate the benefit of using gas rather than coal.[5]

The recent dash to gas in the United States is mainly the result of applying appropriate extraction techniques to reserves caught in shale rock formations, which are inaccessible with traditional techniques and which have proved more and more abundant as exploration has broadened. The techniques for extracting natural gas—namely, horizontal drilling and hydraulic fracturing ("fracking")—were invented in the 1940s and 1950s, but they had remained on the sidelines until the beginning of this century. Since then, these techniques have colonized at neck-breaking speed the most favorable sites, such as the Marcellus Formation centered on Pennsylvania or the Barnett Shale and Eagle Ford Shale (gas and oil in the latter case) in Texas. This rapid increase was made possible by the readily available skills and equipment resulting from the long and sustained drilling activity in the United States. In addition, there has been almost as little federal regulation for fracking as there was for oil drilling when John D. Rockefeller started the industry.

Natural gas drilling has proved to be a bonanza for the U.S. economy, though it comes at several costs. Hydraulic fracturing, as the name indicates, applies pressurized water inside shale rock. Water first travels down the vertical section of the well. When it reaches the level of the target rock, the water stream is redirected along the horizontal section of the well in order to hit and fracture the rock, thus opening pathways for gas circulation. The operation requires sizable quantities of water—anywhere from two to five million gallons for the initial fracking and another two to three million gallons apiece for possible refracking during the well's lifetime. These quantities correspond to the daily consumption of a U.S. town with several tens of thousands of inhabitants. Where water is not scarce, this amount is not significant. Where water is scarce, as in drought-stricken California or Texas, competition with more traditional water users may become tense. And in catchment areas, the effects can be disastrous (Freyman and Salmon 2013).[6]

To boost the fracking effect, the water is loaded with a cocktail of several dozen chemical products, none of which are regulated by the Environmental Protection Agency (EPA), which is responsible for water quality regulation across the United States. Fracking is exempt from such regulation, as are all energy-extraction operations in the United States. Fracking operators are not even systematically required to disclose the composition of their cocktails. The target rocks usually sit far deeper in the ground than the aquifers from which water is pumped for human consumption. Thus, the vertical component of a well goes down through the aquifer, which would raise no problem as long as the casings do not leak—but, of course, some casings do leak.

How polluted water flowing back from fracking wells is disposed of is another cause for concern. The *Texas Tribune* reported convoys of heavy trucks ferrying polluted water to disposal wells in the Texas countryside. These wells are not supposed to communicate with the aquifers, but, in practice, it is impossible to guarantee that there is no path to any of the numerous disused oil wells that go deep underground (Henry and Galbraith 2013). In Oklahoma (and, to a lesser extent, in Kansas and Texas), the indiscriminate disposal of refuse water is responsible for an epidemic of earthquakes. Of course, it would be safer—albeit less profitable for the operators—to treat the polluted water and then recycle it into the fracking process.[7]

Tensions over land use will also become more acute as natural gas drilling spreads.[8] Because the spatial reach of a well is relatively small (much smaller than for conventional wells), the number of wells is large relative to the production achieved, and the density of wells on a given territory is high. The municipality of Fort Worth, Texas, for example, filed a complaint in Tarrant County District Court against Chesapeake Energy, the second-biggest driller in the Barnett Shale of north Texas. The objective, however, was not to have Chesapeake renounce its rights to drill among the city's houses, schools, and public spaces; rather, it was to recoup royalty payments on which Chesapeake had allegedly cheated (and the company has settled). Elsewhere in Texas and other states, conflicts on land use should prove more confrontational. In fact, some local governments in Texas are resolutely opposed to fracking on their respective territories. However, a bill voted on by the Texas state legislature and signed in May 2015 by the governor prohibits those local governments from banning fracking. In Florida, the State House voted on a similar bill; however, in March 2016, the Florida State Senate Appropriations Committee killed the bill by a narrow majority. Indeed, merely considering fracking in Florida's porous limestone soil and fragile aquifers, upon which 90 percent of the residents depend for their drinking water, is sheer madness.

One additional point that is no less important than the preceding ones is relatively rarely mentioned: cheap gas has the potential to displace not only coal but also renewables and nuclear and to weaken whatever incentive might exist to save energy. A study from the Energy Modeling Forum (2013) at Stanford University, after having evaluated at "$70 billion annually over the next several decades" the boost to the U.S. economy, is rather less upbeat about the effects of natural gas on the climate:

> Shale development has relatively modest impacts on carbon dioxide, nitrogen oxide and sulfur dioxide emissions, particularly after 2020. Since 2006, electricity generation has become less carbon intensive as its natural gas share increased from 16 to 24 percent and its coal share decreased from 52 to 41 percent. Over future years, this trend towards reducing emissions becomes less pronounced as natural gas begins to displace nuclear and renewable energy that would have been used otherwise in new power plants under reference case conditions. Another contributor to the modest emissions impact is the somewhat higher economic growth that stimulates more emissions. Reinforcing this trend is the greater fuel and power consumption resulting from lower natural gas and electricity prices.

At the least, one should not exaggerate the effects of the gas business expansion on climate change.

Europe also has significant shale gas reserves—in particular, in Britain, France, and Poland—though their reserves are not as large as those in the United States. It is difficult to imagine that these reserves will be exploited in the same way as in the United States. Population densities are generally higher and environmental sensitivities greater in these areas. In Sussex, Kent, and Lancashire, for example, where the British Tory government has promised that shale formations should be tapped, severe civil unrest is more likely than a Fort Worth kind of approach. Differences in legal dispositions, especially regarding property rights, are also notable. Being owners of the resources beneath their land, many U.S. landowners have willingly participated in the shale gas dash; there are no such incentives in Europe, as underground resources are, in most cases, public property.

In Latin America, the configuration of the shale formations is more akin to the configuration in the United States. This is especially true of the enormous Vaca Muerta Shale southwest of Buenos Aires, near the border with Chile. The geology here is highly favorable, with thick layers of shale, and the area is sparsely populated. The main problem might be in convincing investors wary of Argentina's business climate.

Shale gas reserves in China might be of the same order of magnitude as those in the United States. However, the geology is far less favorable; the terrain is often mountainous, and the underground area is chaotic after millions of years of earthquakes. Moreover, water is generally scarce throughout China and is extremely scarce in some places of interest to natural gas drillers. As for the Chinese operators, which have expressed an interest in drilling for gas, they are relatively inexperienced and lack state-of-the-art equipment. The two leaders, PetroChina and Sinopec, which are also the main state-owned firms in oil and gas, are embroiled in investigations for various alleged wrongdoings—in particular, corruption in their dealings with foreign partners. This makes further procurement of specialized equipment and services problematic, however badly they are needed. Optimistic shale gas production targets for China, announced in 2011–2012, were sharply scaled down by the National Energy Administration in April 2014, dashing hopes that China would have enough gas available for a large-scale coal substitution in the foreseeable future.

Natural gas is often heralded as a bridge from coal to renewables. That bridge, however, should not have too many cracks, and its existence should not impede the development of the very renewables it is supposed to lead to. Otherwise, it will be a bridge from coal to coal (Levi 2013).[9]

OIL

Burning oil is not good for the climate, even if it is less damaging than burning coal. Peak oil might be welcome, but it is not in sight. On the contrary, oil seems to be popping up everywhere, with new discoveries and new technologies that make it possible to extract oil from places that could not be reached before. Even if this trend is frozen by the collapse of oil prices, it remains latent, waiting for prices to pick up.

One such place where oil could not be reached before are shale formations. The United States is again at the forefront of this development, adding weight to the concerns formulated in the aforementioned Stanford Energy Modeling Forum report (EMF 2013). In Pennsylvania, gas is the main product of the shale formations; in Texas, it is gas and oil; and in North Dakota, there is a new frontier of U.S. oil. In fact, operators there disregard and burn on the spot gas coming with the oil. The resulting flares are visible from space and are generous producers of CO_2 emissions.

U.S. shale oil producers are struggling in the current context of prices falling to less than $40 a barrel; in fact, many are heading to bankruptcy. However, those companies that are financially resilient and that own drilling rights in the most favorable areas weather the crisis on the back of continuing productivity gains and will benefit from the eventual price rebound.

The Canadian province of Alberta has been dubbed a second Saudi Arabia, due to the enormity of the oil resources trapped in its tar sands under the form of heavy bitumen. However, exploitation of these tar sands is not profitable as long as the price of oil is under $80 per barrel. Alberta might also be considered a second China, as extracting and refining bitumen and then burning the fuel so obtained is no less damaging for the environment and the climate than the extraction and burning of coal in China. Converting the bitumen into a more convenient liquid consumes about half the energy that will subsequently be obtained from this liquid. Citizens of Detroit, a U.S. city close to Canada, are discovering what it means to harbor a refinery of Canadian oil. One rather unpleasant by-product is a variety of petroleum coke that has a high sulfur content and various other impurities (heavy metals) that make it improper for use in steelmaking. Although it could be burned as a cheap fuel in power plants, this is not allowed in the United States due to the unbearable pollution that is produced. Eventually, this by-product will be exported to some countries in Asia; in the meantime, however, a black mound of petroleum coke along the Detroit River just gets bigger and bigger (Austen 2013). Similar mounds are growing by other northern Midwestern communities, blowing black dust and leaking into watercourses. In Chicago, these mounds pit residents and some of their representatives against Koch Industries, which owns the disposal facility under contract with BP to refine the Canadian bitumen in a nearby plant.

Another factor to keep in mind is that companies and nations are drilling in ever-deeper seas along the African and South American coasts—not to mention the fact that the Arctic reserves are not yet written off. In September 2011, ExxonMobil and Rosneft, a Russian state-controlled oil company, agreed to start a multibillion dollar joint venture to search for oil and gas in the Arctic Sea off the coast of Siberia. There, the waters are not as deep or as rough as the seas off the northern coast of Alaska, where attempts by Royal Dutch Shell have been derailed by tough climatic conditions. Although adverse economic and political circumstances have delayed the materialization of this agreement, they have not killed it.

Thus, there is no shortage of oil, coal, or gas; nor is there a shortage of extremely powerful operators determined to extract the products and to go on pushing them into power plants, vehicles, houses, and so forth. On July 4, 2016, the Oslo-based Rystad Energy consultancy released an evaluation report, titled *United States Now Holds More Oil Reserves Than Saudi Arabia*, of world oil reserves; the report is based on the analysis of 60,000 fields worldwide. The main conclusion reads: "Rystad Energy now estimates total global oil reserves at 2,092 billion barrels, or 70 times the current production rate of about 30 billion barrels of crude oil per year. For comparison, cumulatively produced oil up to 2015 amounts to 1,300 billion barrels. Unconventional oil recovery accounts for 30 percent of the global recoverable oil reserves while offshore accounts for 33 percent of the total." It then issues a warning typical of the fossil fuel industries: "This data confirms that there is a relatively limited amount of recoverable oil left on the planet. With the global car-park possibly doubling from 1 billion to 2 billion cars over the next 30 years, it becomes very clear that oil alone cannot satisfy the growing need for individual transport" (Nysveen 2016). A more relevant warning might have been offered: well before seventy years have passed, the earth's climate will have blown up; only a total ban on coal, severe restrictions on oil and gas consumptions, and carbon capture (see chapter 6) might deflect the trajectory to doom.

RENEWABLE ENERGY SOURCES

Among renewable energy sources, wind, solar, biofuels, and hydroelectricity are already used on significant scales. Wind and solar energy suffer—for now—from a serious intermittency handicap, but that could soon be remedied by advances in energy storage. Moreover, the costs of producing electricity from wind and even more from the sun are declining rapidly. As for biofuels, the ways in which they are currently produced and used are deeply flawed. Dams for the production of hydroelectricity (and, indeed, dams in general) have disturbing, and often devastating, effects on the local populations and on the environment.

The efficiency of both wind and solar energy uses will be boosted by significant improvements in energy-storage techniques, which should be available soon, and by improved demand management. Even for biofuels, there are encouraging prospects with innovative approaches, some of them surprising. These developments will be assessed in chapter 6; here, we consider the present situation.

WIND

Producing electricity from wind is, in favorable places, already competitive with more established energy sources. Although major technological breakthroughs are, in general, not expected (if one ignores, for the time being, futuristic projects like machines flying at an altitude of several thousand meters above the U.S. Midwest to catch strong, sustained currents), significant improvements still prove feasible, especially in terms of the conditions under which turbines can function. (When there is too little wind, they stop; when the wind is too strong, they must be stopped.) In northern Europe, for example, new machines have been erected that can function with winds blowing at speeds as low as 10 m/s. These machines function on average 2,400 hr/yr, instead of 1,800. In the United States, improvements have been made at the other end of the bracket of usable wind speeds, above 25 m/s, allowing up to 4,000 hr/yr. In certain offshore wind farms, newly introduced floating machines might be preferable to anchored ones; the first one—the Hywind floating wind farm by Statoil—is under construction off the coast of Aberdeenshire (Scotland) and should be operational in 2018.

There remain two significant problems with wind energy, despite broader ranges of usable speeds. First, winds are, in general, intermittent and cannot be ordered according to electricity demand. However, temporal imbalances might be smoothed with appropriate high-capacity storage devices and demand management. The second problem is spatial: wind on a large scale is not a local energy. This problem can be illustrated by the geographical imbalance in Germany. Large wind farms are located in the north of the country, where North Sea winds are blowing; however, demand is greater in the south than in the north. Building high-voltage power lines between north and south is fiercely resisted by the populations that might be affected. Texas, on the other hand, doesn't know that kind of embarrassment. Not content to be a heavyweight only in oil and gas, Texas now has vast wind resources and is determined to tap them on a grand scale. Wind power already accounts for almost 10 percent of the electricity generated in the state. But these wind resources are in the west, while the large cities of Houston, San Antonio, Dallas Forth

Worth, and Austin are in the east and southeast. To connect producers with consumers, 3,000 miles of power lines were swiftly built between 2010 and 2013.

SOLAR

Direct solar energy is also renewable and intermittent, though in rather different ways from wind. Solar heat may be concentrated on a boiler located on top of a tower at the center of a field of mirrors. In places where the sun is shining almost all day and all year, as in central Spain, North Africa, or the U.S. Southwest, such solar plants may be competitive with more established ways of producing electricity. When a fraction of the energy collected during the day is used to melt salts that release the heat at night while solidifying, the process is economically more efficient. However, such plants require amounts of water that are not routinely available in the semiarid regions where these plants often exist.

It is also possible to produce electricity directly from solar radiation. Within a photovoltaic cell, photons from solar radiation displace electrons from atoms of an appropriate material; the most commonly used is crystalline silicon, though several others are also in use or under experimentation. This technology is in flux and is in a Schumpeterian phase of development, in which a variety of innovation paths are in competition (see chapter 6). It is thus no wonder that (1) electricity produced from materials that are still relatively primitive is far too costly (2) innovations are driving costs down at a stunning pace.

Some countries have heavily (albeit, prematurely) subsidized electricity produced in onerous conditions, creating misplaced incentives for large numbers of consumers and manufacturers. As a result, a great deal of public and private money is wasted on unsatisfactory equipment, without significant impact on the energy mix and without inducing research and development (R&D) on better approaches. European and U.S. producers of solar panels have not even benefited for long from the fast-growing demand; instead, they have been marginalized, and several among them have been bankrupted by competition from Chinese producers, who are directly and heavily subsidized by municipal and regional authorities and state banks.

Manipulated competition has been so harsh that even some Chinese producers could not survive the reduction of subsidies from disillusioned sponsors.

Public subsidies are justified for the support of R&D efforts, with development being no less important than research; but they are not justified when they distort competition and prematurely disseminate unsatisfactory techniques that, in such conditions, crowd out more advanced substitutes that unduly struggle to enter distorted markets. Fortunately, there are ways out of this situation that have increasingly proved feasible and effective (see chapter 6). If these more successful methods are followed, it will be possible to tap the enormous potential of solar energy at surprisingly low costs.

BIOFUELS

First-generation biofuels (produced from foodstuffs like corn, wheat, sugarcane, and canola) are climatically deceptive, as their production uses up more fossil fuels than their subsequent use in vehicles saves (with the exception of sugarcane in Brazil). They are also socially disastrous, as anybody knows who has seen poor farmers in Central America driven from their plots by agriculture operators and financial groups intent on making money from corn for subsidized fuel. Moreover, they induce land-use changes entailing the destruction of valuable carbon sinks. From investigations reported in *Science*, Fargione and colleagues (2008) concluded:

> Our results demonstrate that the net effect of biofuel production via clearing of carbon-rich habitats is to increase CO_2 emissions for decades or centuries relative to the emissions caused by fossil fuel use. Conversely biofuels from perennials grown on degraded cropland and from waste biomass would minimize habitat destruction, competition with food production and carbon debts, all of which are associated with direct and indirect land clearing for biofuel production. (1237)

Provided they strictly avoid these pitfalls, second-generation biofuels should fare better—in part, because the chemical processes for their production are kick-started and sustained at relatively low temperatures by enzymes functioning as catalysts and, in part, because they do not depend on foodstuffs or on detrimental land-use changes for their production. Biofuels produced from waste wood in forests managed in a sustainable way for timber or pulp production or produced from surplus straw foot the bill; in the same vein, biogas—and heat—from urban and industrial waste is a valuable resource, as demonstrated not only in northern European cities but also in the Milano-Brescia conurbation in Italy. Production of second-generation biofuels can take place where such resources are available, with the enzymes produced where specialized expertise and equipment are concentrated—for example, at the Danish firm Novozymes. However, the advanced processes for producing second-generation biofuels are significantly more costly than the primitive ones for first-generation biofuels. Heavily protected and subsidized (in both the United States and the EU), first-generation producers crowd second-generation ones out of the market—and even out of their fair share of R&D support (Peplow 2014). Other promising approaches (though still at the R&D stage) are considered in chapter 6.

Rather humble compared with these science-based perspectives is a simple biofuel extensively used in some places that is either a real "green" resource or a hoax, depending on its origins. For district heating in Scandinavian cities and for heating private buildings, wood pellets are burned in high-efficiency stoves. Outdated coal-fired power plants have also had their boilers converted at relatively low costs to wood pellets, which is a sustainable practice as long as the pellets are made from marginal forest resources available in the same region and as long as enough vegetal material is left on the ground to regenerate the soil. However, the demand for wood pellets is increasing to such an extent that European forests can no longer meet it in sustainable ways.

It doesn't look rational to square the circle by importing firewood by the shipload from forests on other continents, where it is rarely collected in sustainable ways. Even less rational and less sustainable are imports of palm oil for biodiesel or of corn for ethanol, both of which contribute to the destruction of faraway forests with high carbon content, all for the sake of abiding by quotas of so-called renewable energy. These quotas were set arbitrarily and irresponsibly by EU authorities and certain national governments. The European Commission even ignored warnings by its own scientists (from the Joint Research Center) about the devastating consequences of its biofuel quota policies on Asian rain forests, especially devastating in the peat lands of Indonesia and Malaysia (see chapter 1). It also ignored a recommendation for mandatory carbon accounts, the necessity of which seems obvious for monitoring policies that are cynically sold to European citizens and taxpayers as being driven by a strong concern for the climate.

CIVIL NUCLEAR ENERGY

Many concerns, both real and imaginary, hamper the development of civil nuclear energy. The following two are of particular significance: keeping nuclear power plants safe and avoiding misuse of weapons-grade material. Spent fuel disposal is a third concern of a more long-term character.

Bitter experiences show that nuclear power plant safety cannot be seriously monitored in the absence of a technically competent and institutionally strong regulator that is stubbornly independent of political interference and industry lobbying. It has been tragically exposed that Japan lacked a regulator of that sort. Nor does the United States have one—the U.S. Nuclear Regulatory Commission is beset by issues between commissioners connected with the industry and more independent-minded commissioners. This lack of a strong regulatory body is an even more serious concern now that many U.S. nuclear plants are reaching ages at which difficult and costly decisions must be made—closure or in-depth renovation. These decisions are particularly critical when nuclear plants are close to urban centers, as Indian Point is to New York City. Consanguinity is even worse in Asia. Only a dozen countries (at most) may be considered to have, if not perfect, at least satisfactory regulatory rules and institutions. France is one of those countries, though it didn't come easily; interferences had to be eliminated and weaknesses corrected. Before they were, however, the director of a power plant, obsessed with short-term performances, relaxed the maintenance routines and, for a few months, escaped being spotted. It then took twice as much time to fully recuperate the plant. In Ontario, Canada, similar deviations, on a much larger scale, led to the closure in 1997 of seven reactors (for months to years) run by the utility Ontario Hydro.

Complacency—or worse—on the part of operators and a lack of attention from the regulator may quickly lead to dangerous situations, with lasting consequences even in the absence of any accident. Efficiently monitoring nuclear safety is an extremely demanding job—particularly as the plants get more vulnerable to cyber attacks (Baylon, Livingstone, and Brunt 2015)—that most countries are not able to assume (hence, they should not tinker with nuclear energy). The consequence is that nuclear power plants are niche substitutes for fossil fuel power plants, rather than a global alternative. The word *niche* might sound unduly restrictive—isn't China on course for building dozens of nuclear plants that should be connected to the grid by 2020? Still, even that will amount to a mere 5 percent of the total capacity of electricity production in the country, according to IEA projections. Getting out of the niche would require simpler, cheaper, and safer reactors, but those are not on offer.

Potential nuclear terrorism is another major concern. It is true that it would be easier for would-be nuclear terrorists to steal weapons-grade fissile materials from insufficiently controlled arsenals in some countries that possess weapons, rather than from the storage facilities of utilities and reprocessing plants. However, the use on a significant scale of plutonium in mixed-oxide fuel (uranium and plutonium) is a concern, as it is impossible to assess with complete accuracy the plutonium in circulation. One does not need large amounts to make a crude explosive device.

In October 1977, Ted Taylor, a nuclear physicist, was lecturing a small group from Princeton University about nuclear terrorism at the foot of the World Trade Center in Manhattan. Taylor had been in Los Alamos after World War II as a designer of the largest atom bomb (subsequently rendered obsolete by the hydrogen bomb) and the smallest (the size of a grapefruit) atom bomb, named Scorpio (McPhee 1974). After leaving Los Alamos, Taylor decided to switch his highly inventive mind to renewable energy at the Princeton School of Engineering. But he was preoccupied with what he considered the relative ease of making a crude, but dangerously powerful, atomic weapon that would not be too difficult to manipulate. He was explaining this idea to his small group of colleagues in Manhattan, detailing the rather limited resources in terms of people and equipment (apart from the fissile material) needed to make the device and then to bring down the towers—or at least to engage in a grand blackmail.

That the towers were later brought down by other means should not allow us to dismiss Taylor's demonstration. Nuclear arsenals should be tightly controlled; nuclear energy should be banned from unreliable countries; and, in those countries considered reliable, weapons-grade material should be avoided when composing fuel for nuclear power plants. The particular nature of nuclear energy requires tight precaution that would also have side benefits in terms of safety. Indeed, safety aims not only to avoid accidents but also to minimize the effects of accidents. That the fuel of reactor 3 at Fukushima was made of mixed oxide has been a matter of great concern. What if strong winds had brought south, toward Tokyo, some plutonium dust that had escaped from the exposed fuel rods?

Contrary to a prediction by Glenn Seaborg, the discoverer of plutonium and long-time chair of the U.S. Nuclear Regulatory Commission, in most countries, nuclear energy does not and will not significantly contribute to electricity production. Some expansion is possible, along with strict monitoring, but it doesn't seem possible on a global scale.[10]

4

PERSPECTIVES ON CLIMATE CHANGE

IN 1978, President Jimmy Carter asked the National Academy of Sciences (NAS) for advice on the relevance of results obtained from the recently produced computerized global models of climate change. In response one year later, he received a stern warning:

> The climate system has a built-in time delay. For this reason, what might seem the most conservative approach—waiting for evidence of warming in order to assess the model's accuracy—actually amounts to the riskiest strategy. We may not be given a warning until CO_2 loading is such that an appreciable climate change is inevitable.

After the defeat of Jimmy Carter in the presidential election of 1980, no one in the White House heeded the call until 2015. In 2014, as a kind of echo to the NAS warning in 1979, Henry M. Paulson Jr. (2014), former head of Goldman Sachs and former secretary of the treasury under President George W. Bush, wrote in the *New York Times*: "Waiting for more information before acting is actually taking a radical risk. We'll never know enough to resolve all the uncertainties. But we know enough to recognize that we must act now" ("The Coming Climate Crash," June 21, 2014). In that span of time, thirty-five crucial years have been lost; the United States and most of humankind have stubbornly remained blind to the climate threat, as Europe in the 1930s remained blind to the Nazi threat. The U.S. military, however, has its eyes open. The CNA Military Advisory Board published in May 2014 the *National Security and the Accelerating Risks of Climate Change* report, which warns: "The projected impacts of climate change could be detrimental to military readiness, strain base resilience both home and abroad, and may limit our ability to respond to future demands" (3).

The first section in this chapter offers a short introduction to the history of the greenhouse effect—from Joseph Fourier, who, nearly two centuries ago, understood how such an effect lifted the mean temperature at the earth's surface to the current 15°C/16°C, to contemporary scientists of the earth and atmosphere, who argue that too much of a good thing will have disastrous consequences. It is not possible to allege that warning signs of climate change have been lacking. This chapter reviews three among the most ominous ones, along with the corresponding feedback loop effects on climate change: the accelerating melting of Arctic and western Antarctic ice sheets and terrestrial glaciers, the release of increasing amounts of fossil methane from terrestrial permafrost and the seabed, and an overabundance of extreme meteorological events worldwide.

Warming and acidifying oceans turn against their inhabitants, such as plankton, coral, shellfish, and birds, while rising sea levels threaten human settlements that currently draw great benefits from their proximity to the sea. In addition, destroyed or decaying forests disgorge the carbon they have been storing. In all these ways, immemorial partners are turned into formidable foes. The United States is not spared the effects of these trends; indeed, in many respects, the United States is on the front line of climate change.

In the last section of this chapter, we discuss the rise in carbon dioxide (CO_2) and methane (CH_4) concentrations in the atmosphere and the increase of the earth's mean temperature. As the three figures in the section illustrate, an acceleration in this increase has been clearly visible, particularly during the past five years. What cannot be seen in the figures is the prospect of large nonlinearities—and even abrupt degradation—of irreversible bifurcations toward wild trajectories. As Knutti et al. (2015) recalled: "Large-scale tipping points and thresholds cannot be excluded, for example a dieback of the Amazon rainforest, a shift in monsoon systems, or the release of methane from marine hydrates" (14). That they "cannot be excluded" actually sounds like an understatement.

THE GREENHOUSE EFFECT: FROM JOSEPH FOURIER TO JAMES HANSEN

In an 1827 paper dealing with the energy equilibrium of the planets, French mathematician and physicist Joseph Fourier explained how a greenhouse effect regulates the earth's temperature: a fraction of the incoming solar radiation is reemitted from the earth's surface to the atmosphere under the form of outgoing infrared radiation (what Fourier called *chaleur obscure*, or "invisible heat"); gases in the high atmosphere then interact with this infrared radiation and trap a fraction of it. Without that effect, the mean temperature of the earth would have been −18°C, instead of +15°C.

Forty years later, the Irish physicist John Tyndall identified three gases in the atmosphere (we now call them greenhouse gases, or GHGs) that are responsible for the bulk of the greenhouse effect: transient water vapor, long-lived CO_2, and much less long-lived but far more powerful CH_4. Why these gases, and not oxygen and nitrogen, which are far more abundant in the atmosphere than CO_2 and CH_4? Seventy-five years later, quantum mechanics provided the clue: GHG molecules oscillate at frequencies that enable them to absorb and reemit parts of the infrared radiation spectrum (for example, wavelengths above 12–13 micrometers [μm] for CO_2).

The Swedish chemist Svante Arrhenius, working at the beginning of the twentieth century, gave the first quantitative estimation of the greenhouse effect—that is, an estimation of the effect that increasing concentrations in the atmosphere of the gases identified by Tyndall would have on earth's temperature. His predictions are, partly by chance, in line with the results from present-day computerized climate models. The first two such models appeared around 1970—in particular, a model produced by U.S. climatologist James Hansen and his team at the NASA Goddard Institute for Space Studies. These are the models that prompted President Carter's inquiry.

During the first half of the twentieth century after Arrhenius, there was a long period of inactivity, and even indifference, bred by a somewhat offsetting effect from atmospheric pollution in the lower atmosphere—in particular, sulfur dioxide (SO_2) from volcanoes and industry. That period ended with the decision by two U.S. scientists—oceanographer Roger Revelle and geochemist Charles David Keeling—to build an observatory for measuring CO_2 concentrations; they chose a place relatively free of local interference, Mount Loa in Hawaii. The main output of their observations was the Keeling curve (see figure 4.1 at the end of this chapter), which started in 1957 at a concentration of 314 ppm (parts per million) of CO_2 in the atmosphere (it was about 290 ppm in 1880) and climbed at an accelerating pace to 380 ppm in 2005 (the last measurement before Keeling's death). These data were both an incentive and an input for climate modelers; they also informed a report that Revelle handed to President Lyndon Johnson in 1965. Thirteen years later, President Carter had a serious look at the science of climate change, but the issue was neglected by his successors until President Barack Obama made it a priority during his last two years at the White House.[1] Since 1988, the Intergovernmental Panel on Climate Change (IPCC) has been leading an effort, at a scale without precedent, to make available all the scientific results that could help improve the understanding of climate change and the necessary modalities of action.[2]

WARNINGS AND FEEDBACK LOOPS

CANARY—NOT IN THE COAL MINE, BUT AT THE POLES

> Of the people involved in global warming, I think we're on top of the list of who would be most affected. Our way of life, our traditions, maybe our families. Our children may not have a future. I mean, all young people, put it that way. It's just not happening in the Arctic. It's going to happen all over the world. The whole world is going too fast.
>
> —John Keogak, Inuit hunter, quoted in Kolbert (2006)

The National Snow and Ice Data Center (NSIDC), located in Boulder, Colorado, has been monitoring the condition of the ice in the Arctic since 1976, when it took over from the World Data Centre for Glaciology. Each September, NSIDC estimates the lowest extent of the Arctic sea ice for the year. The minimum of the lowest extent was reached in September 2012, with 3.41 million km^2, or half the 1979–2000 average. September 2014 was rather less extreme, with 5.30 million km^2. September 2015, which reached 4.41 million km^2, reset a downward trend that was confirmed by a 2016 level close to the 2012 one.

These reductions in extent reinforce a feedback loop: the more water takes the place of ice, the more the already strong warming trend in the Arctic (up to +6°C right above the surface of the sea) is reinforced. This is because water reflects rather little (less than 10 percent) of the solar radiation, whereas ice reflects more than 60 percent. This difference in albedo, i.e., the proportion of the incident radiation that is reflected, is considerable; hence, the increasing quantity of heat absorbed by the sea. The feedback to climate change is thus considerable as well.

There will soon be another feedback loop that is less direct but still significant. Most countries around the Arctic Ocean have made clear their intentions to make the most of easier access to the resources that might be extracted from the seabed—in particular, oil and gas. Indeed, Russia is determined to have its "sovereignty" recognized there. Burning fossil fuels is the main cause of climate change and hence of Arctic ice decay; now this very decay opens the way to accessing more stuff for burning. It should also be noted that part of the radiation absorbed in the summer comes out in the winter as heat, changing the circulation patterns in the atmosphere—notably, the jet stream—in complicated ways that affect the weather at lower latitudes in Europe and North America.[3]

The extra heat in the Arctic Ocean significantly contributes to melting the adjacent Greenland ice sheet (Mouginot et al. 2015). The same process of eating from underneath at the ice is being observed in western Antarctica, also at an accelerating pace. On May 12, 2014, NASA released a joint communiqué with the University of California at Irvine, titled "NASA-UCI Study Indicates Loss of West Antarctic Glaciers Unstoppable." The study is based on observations gathered from 1992 to 2011 in the Amundsen Sea sector of the Southern Ocean, near western Antarctica, where a vast range of glaciers is retreating. According to the communiqué, the conclusions of the study are stark: "This group of glaciers release almost as much ice as the entire Greenland Ice Sheet. At this point, the end of this sector appears to be inevitable." (For a scientific presentation of the Study, see Rignot et al. 2014.) It is now seriously considered that western Antarctica alone might contribute 1 m to sea level rise by 2100. For now, the enormous mass of ice covering eastern Antarctica does not show signs of weakness, contrary to Greenland and western Antarctica.

A few years ago, it was not envisaged that the melting of the ice sheets in Greenland and western Antarctica would, by the end of the

BOX 4.1 GLOBAL WARMING POTENTIAL

Global warming potential (GWP) is a concept introduced by the Environmental Protection Agency (EPA) for assessing the greenhouse potential of a gas. It is a measure of how much energy 1 ton of a gas emitted will absorb over a given period of time, relative to 1 ton of CO_2. The GWP of CH_4 is 72 over 20 years and 21 over 100 years. The difference between 72 and 21 reflects the fact that the average lifetime of CH_4 in the atmosphere is 12 years, whereas CO_2 remains in the atmosphere as long as it is not absorbed by oceans, terrestrial vegetation, or artificial devices.

Other GHGs are emitted in significantly smaller quantities but have larger—and, for some, far larger—GWPs. Nitrous oxide (N_2O) has a GWP in the range of 265 to 300 for a 100-year timescale, whereas chlorofluorocarbons (CFCs) and hydrofluorocarbons (HFCs) have GWPs in the thousands or tens of thousands. About two-thirds of the anthropogenic N_2O emissions result from the application of huge amounts of nitrogen fertilizers in industrialized agriculture. CFCs are used in refrigeration and air-conditioning equipment and as propellants and solvents, though they are being phased out under the 1987 UN Montreal Protocol on Substances That Deplete the Ozone Layer. HFCs looked like convenient substitutes before their high GWP was recognized. According to the October 2016 UN Kigali Amendment to the Montreal Protocol, they will be phased out starting in 2018 in developed countries, albeit only in 2024 in China and 2028 in India—delays that might result in blown-up emissions.

century, add more than 50 cm to average sea level; now, somewhere between 1.0 and 1.5 m looks more realistic, according to the glaciologists monitoring the region. As Dr. Gordon Hamilton, glaciologist at the University of Maine, noted in the *New York Times*: "All these changes are happening at a far faster pace than we would have ever predicted from our conventional theories" (Justin Gillis, "As Glaciers Melt, Science Seeks Data on Rising Seas," November 13, 2010).[4]

In the future, the pace might hang on what threatens to become the most potent feedback loop: as permafrost in the Arctic melts, methane is released. The total quantity of methane actually released is still modest compared with CO_2 emissions from burning fossil fuels, but that might change as enormous reserves of methane trapped under shallow waters are tapped by climate change and come to reinforce it.

Eric Kort, from NASA's Jet Propulsion Laboratory in Pasadena, California, and his team have detected at the surface of the Arctic Ocean unusually high CH_4 concentrations that suggest it is released in increasing quantities from underwater sources (Kort et al. 2012). Igor Semiletov, while leading the joint Russia-U.S. effort meant to explore the less well-known Siberian portion of the Arctic sea, identified credible culprits: enormous plumes of methane bubbling from the sea. In an interview with *The Independent*, Semiletov recalled how surprised he had been when discovering these plumes:

> I was most impressed by the sheer scale—one km or more wide—and the high density of the plumes. Over a relatively small area we found more than 100, but over a wider area there should be thousands of them. . . . One of the greatest fears is that, with the disappearance of the Arctic ice in summer, and rapidly rising temperatures across the entire Arctic region, the trapped methane would be swiftly released into the atmosphere, leading to rapid and severe climate change. (Interview with *The Independent*, December 13, 2011)

HIGH MOUNTAIN GLACIERS: THE GREAT RETREAT

In 2005, Dr. Edson Ramirez, a leading Bolivian glaciologist, was considered a pessimist when he warned that the Chacaltaya glacier (elevation 5,421 m) would most probably disappear by 2015. Chacaltaya is not a large glacier, and its disappearance should not significantly reduce the availability of water in La Paz and its poorer sister city, El Alto. However, for people living there, the glacier *was* a friendly neighbor (about 20 km from the city)—it had disappeared by 2009.

Chacaltaya was at the vanguard of the Andean tropical glaciers that tower over Bolivia, Ecuador, and Peru, and most of them are relentlessly losing ice. Before 1975, they were more or less stable; from 1975 onward, however, they have lost about 40 percent of their volume of ice. Their complete disappearance is expected to occur sometime between 2025 and 2040, if not earlier if the process accelerates, as it did for Chacaltaya. It is not only the poignant beauty of the Andean landscape that is at stake. During the dry season, extending from April/May to September/October, rainfall is very limited and is becoming even more so. During Bolivia's dry season, inhabitants of La Paz and El Alto, as well as the Aymara and Quechua people living on the Altiplano, depend on the flow of water regularly released by the Tuni Condoriri glaciers. Glaciers also store extra water during wetter years that is most welcome during dryer years.[5] The hydroelectric power plants that provide up to 40 percent of the electricity consumed in Bolivia depend on the water from two other glacier ranges—Charquiri and Zongo. As these glaciers melt, extra water is made available for uses downstream, though these waters are not welcome in times of flooding during the wet season. As the volumes of ice decrease, so do the volumes of water flowing from the mountains during the dry season.

Due to demographic pressure (within thirty years, El Alto has grown from a relatively small *población* next to La Paz airport to a city-slum with a population of nearly a million people), poverty, and mismanagement, water is already scarce in La Paz and El Alto; it is difficult to imagine how these cities will cope with significantly decreasing resources. As for the Indian peasants on the Altiplano, they will have no other choice but migration—but where to? (Oxfam International 2009). The semiarid coastal plains in Peru, where Lima (population: 8.5 million) sits, are at least as dependent on glacier water as are La Paz and El Alto. The prospects there are thus no better (Painter 2008).

All Andean tropical glaciers are highly sensitive to climate change. As a report by the Committee on Himalayan Glaciers, Hydrology,

Climate Change, and Implications for Water Security (2012) makes clear, there is no such uniformity in the evolution of Himalayan glaciers, which are situated at more northern latitudes and are part of a far more extended range of mountains. On the basis of a broad review of available studies on the subject, the report concludes that glaciers in western Himalaya (which depend on winds blowing from Europe, called *westerlies*, for their winter recharge) are, in general, stable or even growing, whereas glaciers in eastern Himalaya (which depend on the Indian monsoon for their summer recharge) are retreating at an accelerating pace as the regional climate becomes warmer and drier. The higher the altitude, the faster temperatures rise.

The glaciers on the eastern fringes of the Himalayas, which are the headwaters to the Yellow River, the Yangtze, the Brahmaputra, and the Indus, have been particularly hard hit because they are tainted by black carbon (soot) brought by strong winds from burning forests and peat bogs in Southeast Asia (see chapter 1) and from heavy air pollution in China. This pollution significantly decreases their albedo. The acceleration of the loss of ice and the effects on the Yellow River and the Yangtze are documented in the *Second Inventory of Chinese Glaciers*.[6]

Currently, in the Yellow River Basin, the per capita water availability is a mere quarter of the national average, which is modest compared with many developed and emerging countries. As economic development continues unabashed in the basin, ignoring the inadequacy of water resources (see chapter 2), no-flow events along the Yellow River get more frequent and extended. Without the water provided by the glaciers, such events would become the norm during dry seasons. In addition, with the reduction of storage capacities in the mountains, devastating floods during wet seasons would become an even more serious threat downstream than they already are today. Bangladesh, India, and Pakistan are under similar threats.

The situation in Central Asia is less well known but no less worthy of attention. From Tashkent in Uzbekistan east-northeastward to Ürümqui in China (Xinjiang), the Tian Shan ("Celestial Mountains"), a range of mountains culminating at 7,430 m, stretches 2,500 km. The Tian Shan glaciers feed the lowlands of Kazakhstan, Kyrgyzstan, and Uzbekistan, where the semiarid climate makes agriculture dependent in the dry season on meltwater for irrigation; it is, in fact, one of the regions in the world most dependent on irrigation. As in the eastern Himalayas, there is no glacier recharge in winter, when it is too dry and cold. Recharges instead depend on the amount of snow falling in summer; with summer temperatures significantly increasing, however, that amount is steadily decreasing. It is no wonder that the glaciers are shrinking. From 1961 to 2012, they lost 20 percent of their surface area (about 3,000 km^2) and 30 percent of their mass. The anticipated losses from now to 2050 are of the order of 50 percent of the mass of ice still existing (Sorg et al. 2012). It is difficult to imagine how a rapidly increasing population will cope with such losses.

EXTREME METEOROLOGICAL EVENTS

Only recently has the hand of climate change been firmly identified in the genesis of extreme meteorological events, along with weather variability. The transition from skepticism to recognition is grounded in a series of events roughly spanning 2005 to 2012. In this section, we discuss a process of investigating and analyzing the events and then apportioning responsibilities among possible causal factors using newly developed statistical tools.

INDIAN MONSOONS

In the past, Indian monsoons displayed a significant amount of variability from year to year, albeit with some regularity over long periods. From 1930 to 1970, there was more above-average than below-average annual rainfall, and floods were relatively frequent. From 1970 to 2000, the reverse was recorded, and droughts were relatively frequent. Since 2000, there doesn't seem to be a clear trend, though there is a greater occurrence of extreme events, often "striking at an odd time or in the wrong place," according to an Indian delegate at the 2006 Earth System Science Partnership meeting in Beijing. He gave the following illustration: "A late onset of monsoon rains in the Marathwada region of Maharashtra state this year caused a mix-up that resulted in 400 drought-struck villages being wiped away by flood waters." He concluded, "One must understand that this is a new ecological order. The planning should be shifted" (Jones 2006).

Devastating floods have also become recurrent in Pakistan during the monsoon. In 2010, the floods were of such extraordinary proportions that they overshadowed those in 2011, which remained largely ignored outside Pakistan, despite over 5 million people being affected. Also in 2011, the monsoon terminated in Thailand with heavy—though certainly not unprecedented—rains that translated into exceptionally devastating floods. With these floods, the critical factors are neither meteorological fluctuations nor climate change; they proceed from inept land use and, in the case of Bangkok, excessive pumping in the underlying aquifer. Flawed land use planning and implementation often turn an extreme event into a catastrophe.

After two years of insufficient monsoons, in 2014 and 2015, the surface water reserves in India and Pakistan were depleted. The drought and heat wave that hit a third of the subcontinent during the 2016 spring were thus all the more devastating (see chapter 2).

DROUGHTS IN KENYA

From 2006 to 2011, the northern part of Kenya was hit by recurrent droughts. The climate there is indeed semiarid, and droughts are not uncommon, but droughts with such a persistent intensity are. The lack of water, compounded by acute problems such as poor management of the water supply, an influx of refugees fleeing the wars in Somalia, and population explosion (a thirtyfold increase since 1900), made life almost impossible in the region. As explained by Anemoi Lokodi, leader of a tribe comprising about 4,000 people who previously made their living by raising cattle: "For us it is like a curse. We have never witnessed such a situation since we were born."

DROUGHTS IN THE MURRAY-DARLING BASIN, AUSTRALIA

The Murray-Darling Basin is a 1 million km^2 water catchment stretching through four Australian states: South Australia, Victoria, New South Wales, and Queensland. It is, by far, the most productive agricultural region in Australia. Since the beginning of this century, it has been hit by recurrent droughts that have been ranked as the worst ever since records began in 1891. Droughts are familiar to farmers in the Murray-Darling Basin; for years, those farmers considered that the current droughts were just another manifestation of weather fluctuations like El Niño and La Niña. However, the recurrence of the phenomenon, the intense heat, and the devastating bushfires that ensued prompted the search for additional explanation and for sweeping reforms in the way water is managed and used throughout the basin (see the Introduction to this book). Weather fluctuations obviously play their part in the emergence or continuance of the droughts, but so do relatively new modes of atmospheric circulation linked to climate change—for instance, more energetic currents above the South Sea that pump humidity off southern Australia.

Heat Wave and Drought in Russia

In a large part of western Russia, including Moscow, July 2010 was the "hottest ever recorded," according to the Russian National Centre for Hydrodynamics. Persistent drought and ubiquitous wildfires accompanied the heat wave. Poor roads and weak public services hindered the efforts to mitigate the disastrous effects. According to the reinsurance company Munich Re, the largest one worldwide, at least 56,000 people died from the effects of heat and smoke emanating from gigantic peat and forest fires. The grain harvest was badly reduced, by about a third, leading the government to ban exports of grain, resulting in yet another spike in the prices on the world market. Another heat wave hit Russia in 2012; though it was less intense, Siberia suffered from numerous and destructive forest fires. Other countries in Eastern Europe and in the Middle East have also been hit by droughts, though not on the same scale.

Drought and Then Floods in Mexico and Central America

Drought and then floods left no respite for farmers in central Mexico, Honduras, and some other areas in Central America during spring and summer 2011, making agriculture there even more precarious than usual. A farmer in Tlaxcala (central Mexico), interviewed by an investigator of the EU project Environmental Change and Forced Migration Scenarios (EACH-FOR), made this sober comment: "The rain is coming later now, so that we produce less. The only solution is to go away to the U.S., at least for a while" (Alscher 2010, 182).

Heat Waves and Droughts Throughout the United States

"Texas Drought 2011: State Endures Driest 7-Month Span on Record" reported the *Huffington Post* on September 5, 2011. In Texas, heat and drought have indeed been extremely severe. Austin, for instance, sweated through ninety days of more than 100°F (38°C) temperatures. It is not disputed that La Niña played a significant role in generating this extreme weather; however, the 2011 La Niña episode was only the sixth strongest since rigorous records started in 1949.

However severe that period of extreme weather was, it pales in comparison with what the following years had in store. During spring and summer 2012, Texas again suffered through a heat wave and drought of historical proportions, though it was just one victim nationwide that year. On August 1, 2012, the number of counties listed as a USDA "primary disaster area" increased by 218 to reach 1,584 counties in thirty-two states. This was in recognition of the effects of a drought considered to be the worst in decades. The geographical extension was exceptionally large, as was the number of activities affected. Where the soils are clay-rich (eastern Texas, for instance), they shrank, and highways cracked. In Midwestern states, highways tended to expand beyond their design limits and then pop up. In many places, power plants had to be shut down for lack of cooling water, and increased demand of electricity for air-conditioning could not be met. Agriculture deeply suffered. Approximately one-third of the world's corn and soybean crop is usually harvested in the Midwest; the persistent and extreme drought reduced the harvests such that they sent prices on the world market soaring.

Two years later, California, parts of Texas, and several other southwestern states were still in the grip of severe drought. In California, according to Park Williams et al. (2015), "Anthropogenic warming has intensified the recent drought as part of a chronic drying trend that is becoming increasingly detectable and is projected to continue growing throughout the rest of this century." The drought there looked endless, before a particularly strong El Niño episode developed in autumn 2015. Despite its strength, this episode didn't provide more than an intermission; there has been another intermission during the 2016–17 winter, and more to come, in what will remain a "drying trend."

Apportioning Responsibilities

To what extent are extreme meteorological events due to either climate change or weather variability? How to apportion responsibilities?

Relying exclusively on statistical investigations of real world data, hence not on climate models, Hansen, Sato, and Ruedy (2012) came to the following conclusion:

> An important change is the emergence of a category of summertime extremely hot outliers, more than three standard deviations (3σ) warmer than the climatology of the 1951–1980 base period. This hot extreme, which covered much less than 1 percent of Earth's surface during the base period, now typically covers about 10 percent of the land area. It follows that we can state, with a high degree of confidence, that extreme anomalies such as those in Texas, Oklahoma, and Mexico in 2011, and in a larger region encompassing much of the Middle East and Eastern Europe in 2010, were a consequence of global warming because their likelihood in the absence of global warming was exceedingly small. (E2415)

This approach is corroborated by the broad methodological effort that has produced a National Academies of Sciences, Engineering, and Medicine (2016) handbook, titled *Attribution of Extreme Weather Events in the Context of Climate Change*.

Although it is not possible to apportion causes for all extreme events, this is not a reason to doubt the increasingly significant role of climate change. In a comment on the subject, listed in the online edition of *Nature* (February 16, 2011), Dr. Donald Newgreen offered an analogy that looks helpful: "I doubt if any individual's lung cancer can be absolutely attributed to their two pack a day smoking habit. But does that disprove the contribution of smoking to the development of lung cancer? What can be done is to ascribe a probabilistic role. The same will hold true with extreme weather conditions."

FRIENDS TURNING FOES: OCEANS, FIELDS, AND FORESTS

LIFE IN THE OCEANS

Oceans absorb nearly 30 percent of CO_2 emissions and 90 percent of the excess heat produced during the warming process that dramatically accelerated in the second half of the twentieth century. In these ways, oceans have hidden from humankind the full significance of the warming trend. However stable these figures might look, they are not physical constants; as the process amplifies, saturation effects shall bear on them. And above all, they come at a high price for the diversity of life in the oceans (Gattuso et al. 2015).

Water warming has major effects on life in the ocean. Some of these effects anticipate the "future global redistribution of marine biodiversity," which is assessed in great detail in Molinos et al. (2015), who remind us: "Current biodiversity patterns have been established well before the Pleistocene, over 2.5 million years ago. Our projections, however, suggest strongly that anthropogenic climate change will drive generalized changes in the global distribution of marine species over the course of a century" (87). The regional variability of these changes will be considerable, with tropical waters and even more equatorial ones suffering large net losses, imperiling the survival of numerous coastal communities.

Some of the following recent changes illustrate what is in the offing:

- In 2015, along the coasts of Maine, lobster populations boomed in moderately warmer waters; in southern New England, at the same time, lobster populations plummeted to record lows in waters that had become too warm.

- The sex of sea turtle offspring is determined by the temperature of egg incubation in sandy beaches; it appears that on more and more beaches, incubation temperatures are above 29°C, the threshold beyond which offspring are female only (Hawkes et al. 2007).
- During the summer of 2015, guillemots along the coast of Alaska (where they are also called *murres*) failed to breed and started to die by the tens of thousands. The most likely explanation is that they were starving, as the marine food chain had been severely disrupted by two years of extreme temperatures in the northern Pacific, driving the creatures upon which guillemots feed to unattainable depths (Atkinson 2017).

Acidification of seawater increases as CO_2 accumulates, reducing the availability of dissolved carbonate minerals (such as aragonite and calcite) required for the building of skeletons or shells. In turn, corals, oysters, clams, mussels, pteropods, and so forth must spend more energy to extract these minerals, hampering their growth and, past a certain point, their survival. A pteropod is a tiny snail protected by a kind of crystal shell; billions of these snails live in relatively cold waters, where they make a staple diet for large fishes like cod, pollack, and salmon. Excessive acidification makes them unable to build their shells and even dissolves existing shells; this is already happening off the northern Canadian coasts, where acidification is enhanced by ice melt. On the other side of the planet, in the Southern Ocean, acidification attacks the calcifying plankton (coccolithophores), which not only is the origin of the food chain but also "plays a key role within the global carbon cycle" (Freeman and Lovenduski 2015). The Southern Ocean absorbs an unusually high fraction of CO_2 emissions; in turn, the shells of the plankton sink, thus transferring carbon from the surface to the deep sea.

Oysters suggest another scenario for *It Happened Tomorrow* (see chapter 1). The seawater along the U.S. northwest coast is naturally among the world's most acidic (U.S. Global Change Research Program 2014); thus, the current acid content of this water, which is being increased by climate change acidification, prefigures general future acidification. So does the condition of oysters there: "We have a nursery where we set oysters continually, but now they can't develop a healthy shell," laments Paul Taylor, the largest shellfish supplier in the Pacific Northwest (quoted in Davenport 2014).

Other living organisms, from unicellular up to fishes, are affected by acidification because they have to spend more energy to maintain the acid-base balance of their inner fluids. This requirement negatively interferes with the functioning of their immune system, as well as with their growth and reproduction processes; for certain species, it even hampers their olfactory and auditory abilities.

Coral reef collapse destroys a precious gift from the sea and is worth special attention. The world's coral reefs are famous for their beauty, but they are no less remarkable as ecosystems harboring a wealth of marine species, including fishes that provide food resources essential to coastal populations on tropical shores. Unfortunately, they are particularly vulnerable to some of the ills plaguing oceans:[7]

- *Warming of the surface waters*: In 1998 and in 2010, which were particularly warm years with El Niño episodes, approximately 15 percent of the world's shallow-water reefs died. Many more were bleached but somehow recuperated. Climate change induces a relatively slow—albeit, continuous—increase in sea temperature that, in turn, increases the potential of natural meteorological fluctuations for devastating coral reefs. This is another manifestation of the "springboard effect." As Mark Eakin, head of NOAA's Coral Reef Watch program, put it: "It is a lot easier for oceans to heat up above the coral thresholds for bleaching when climate change is warming the baseline temperatures. If you get an event like El Niño or you just get a hot summer, it's going to be on top of the warmest temperatures we've ever seen" (quoted in Gillis 2010).
- *Cyclones and hurricanes*: On the Great Reef Barrier and in the Caribbean Sea, these storms are a significant factor of coral destruction. Cyclones and hurricanes are natural meteorological events; however, the potential of their destruction is enhanced by climate change.
- *Seawater acidification*: As seawater gets more acidic, the availability of carbonates, which, as already mentioned, are necessary for building coral skeletons, severely declines.

For all of these reasons, coral formation is no longer able to keep up with natural and accidental decay. Callum Roberts did not paint an unduly bleak picture of the situation when he wrote in his book *Ocean of Life* (2012): "It is chilling to think that, within the space of 100 years, humanity could reverse a process of coral reef formation that has flourished since the end of the last ice age" (107).

COASTAL HUMAN SETTLEMENTS

Synthesizing several studies on the perspectives of sea level rises linked to climate change, Overpeck and Weiss (2009) concluded: "It would be wise to assume that global sea level rise could significantly exceed 1 meter by 2100 unless dramatic efforts are soon made to reduce global greenhouse-gas emissions" (21462). This prediction is in line with the anticipations made by glaciologists monitoring Greenland and western Antarctica (see earlier in this chapter). Indeed, having developed a model that more fully integrates the potential contribution of Antarctica, DeConto and Pollard concluded that "Antarctica has the potential to contribute more than 1 meter to sea-level by 2100" if CO_2 emissions are not drastically reduced (DeConto and Pollard 2016, 591).

Which communities would suffer most from such rises? The answer is those with large coastal populations and high concentrations of activities in low-lying areas. Rising sea levels and stronger storm surges (here, again, the "springboard effect" is at work) would have particularly devastating consequences in large deltas (the following list is illustrative, not exhaustive):

- *Ganges-Brahmaputra-Meghna delta in Bangladesh*: This delta extends through approximately one-third of the country and harbors two-thirds of its population. Tens of millions of people would be terminally, or at least repeatedly, flooded; many would have to leave because of the destruction not only of their dwellings but also of their livelihoods.
- *Mekong delta in Vietnam*: This region is already highly vulnerable to storms and high waters. It is also threatened by the industrialization of the country and by dams upstream in China and Laos,

which prevent sediments from reaching the delta, where they play a vital role. Significant sea-level rise, with present land elevation at less than 1 m in many places and less than 4 m all over the delta, might be a terminal blow for the twenty million inhabitants concerned, as well as for the rice-growing and fishing activities that provide half the food produced in Vietnam.

- *Nile delta in Egypt*: The Nile delta is threatened not only by flooding but also by salinization (already on the rise) of ground and surface waters, which is all the more difficult to contain because the river flow is very low and polluted once the Nile reaches the delta. Approximately twenty million people are directly concerned, one-third of them younger than fifteen years of age. The delta is Egypt's garden, providing more than half the fish captured in the country.
- *Meuse-Rhine delta in the Netherlands*: In the Netherlands, the land (with one of the highest densities of population and activities in the world) is so low that it would have long been submerged but for an elaborated system of dykes and surge barriers. If that system is not suitably enhanced in step with the changing circumstances, most of the country will disappear. It seems that it will indeed be suitably enhanced, as the example of Rotterdam shows: on the ground and on the water, the city and the port prepare for the coming storm surges with a comprehensive mix of technology, cooperation with nature, urbanism, and social mobilization. It looks like a model for other coastal cities.
- *Mississippi delta in the United States*: In this region, storm surges are particularly strong. Without a defense system such as that of the Netherlands (but designed in harmony with the natural characteristics of the Gulf Coast and the Mississippi River—see chapter 6) people and industrial activities would have to be moved elsewhere.

In some places, sea level rises more than elsewhere; in other places, land shrinks, though with the same effect of increased vulnerability. The cause of this shrinking land may be natural subsidence; however, it is often caused by humans, the result of compaction under the weight of buildings or of excessive withdrawals from an underlying aquifer. Large cities like Bangkok, Jakarta, Mumbai, and Shanghai are prominent examples.

Finally, we should be aware that more coastal communities are under serious threat than those mentioned above, and their number keeps rising. In addition, the dynamics of climate change is such that its effects will keep amplifying for a long time. Hence, the coastal communities "will have to continually resettle on higher ground. The stability of the conditions that made the development of those communities possible and profitable will be gone forever.

FIELDS AND FORESTS

More CO_2 in the air means more plant growth. If nothing else changes in the plant environment, it is indeed true that photosynthesis works in this way. However, as a more general proposition, this statement is not true, contrary to what enthusiasts of CO_2 build-up in the atmosphere suggest. Simultaneously changing some significant parameters of the plant environment—in particular, temperature and water availability—radically changes the perspective.

Arabica coffee cannot be grown outside places where specific conditions of both soil quality and temperature are met. According to a study sponsored by the Kew Royal Botanic Gardens (Davis et al. 2012), at least 65 percent and possibly up to 95 percent of such places will have disappeared by the end of the century. Admittedly, Arabica coffee is a demanding plant; however, less extreme but more general findings point in the same direction: "Warming has already lowered yields of wheat and corn" (Jones 2011). As summed up by David Lobell (2011), from the Center on Food Security and the Environment at Stanford University:

> Progress in crop science has shown that most crops show fairly rapid declines in productivity as temperatures rise above critical thresholds, with as much as 10 percent yield loss for +1°C of warming in some locations. Smallholders—coffee farmers in Peru, tea growers in Tanzania, rice growers in India . . .—have already suffered severe yield reductions that threaten their very survival. Both sub-Saharan Africa and South Asia appear particularly prone to productivity losses from climate change, in part because major staples in these regions are often already grown well above their optimum temperature. (2)

Confronted with both dramatically increased population and significantly decreased food availability, how will humankind be able to cope? Even stemming all sources of waste in food production, processing, distribution, and consumption might not be enough.

During the summer of 2003, a heat wave (meaning several consecutive weeks above 30°C) hit Western Europe. Trees such as oaks and beeches reacted by cutting themselves off from their environment such that they drastically reduced their water requirements. In so doing, they also cut their CO_2 intake, and hence their growth, as could later be seen from the unusual thinness of their rings for that year.

Heat and drought during the summer and part of the spring are now the norm in the American west, from Arizona and Texas up to British Columbia (see earlier discussion). Even Alaska is not spared. Weakened trees are then prey to insects and wildfires that are "extraordinary in terms of their severity, on time scales of thousands of years" (quoted in Gillis 2011).[8] Similar trends are observed in the immense Siberian forest and in rain forests—particularly, the Amazon—with unwelcome regional effects on water cycles, for instance. There are also the global effects on the climate that occur as forest is taken over by savannah. Forests, vegetation, and soil (to which carbon has been transferred from the atmosphere by the trees) combined contain at least twice as much carbon as the atmosphere, and the Amazon forest is by far the largest rain forest in the world (see chapter 1). That is a lot of carbon to be released through organic decay and fires as forests turn from friends to foes.[9]

IS THE UNITED STATES ON THE FRONT LINE OF CLIMATE CHANGE?

It is generally recognized that many countries in Africa, the Andean countries, China, India, and the Philippines (to name a few) will pay a particularly high price for climate change. Year after year, the Philippines have been struck by ever more destructive typhoons. In India, "centennial" rains in autumn of 2015 devastated the city of Chennai and its surroundings. In addition, extreme drought in the spring of 2016 led many hundreds of desperate farmers to commit suicide (see chapter 2). Having squandered its natural capital, China shall struggle more and more to cope with the effects of climate change.

Is the prosperous United States also on the front line? Some recent papers and reports suggest that the question should not be lightly dismissed. Jones et al. (2015) found that "U.S. population exposure to extreme heat might increase in the latter half of the twenty-first century four- to sixfold over observed levels in the late twentieth century" (652). Contrary to common perceptions, the consequences might be even worse in the eastern United States than in the west, as "the combination of high temperatures with high humidity is significantly more uncomfortable, and potentially more dangerous, than high temperatures under drier conditions" (Hsiang et al. 2014).[10]

In such conditions, the standards of public health that Americans generally take for granted are not sheltered from the predictions in the 2015 report of the *Lancet Commission on Health and Climate Change*: "The effects of climate change are being felt today, and future projections represent an unacceptably high and potentially catastrophic risk to human health" (Watts et al. 2015, 1861). Such risk is systematically assessed in a scientific report sponsored by the U.S. federal government (Crimmins et al. 2016). In addition, one should also consider the following:

- The Atlantic coast is more vulnerable to rising sea levels than most coastlines in the world, due to specific conditions of water currents, air pressure, and ground subsidence. Places like Annapolis, Maryland; Norfolk, Virginia; Charleston, South Carolina; and Fort Lauderdale and Miami Beach, Florida, suffered from six to eleven times as many days of tidal flooding per year over the past twenty years as they did before then. In fact, parts of historical Charleston are now flooded for an average of 220 days per year.
- Stronger and more frequent floods are also expected, as documented in successive *National Climate Assessments.* The devastating floods in Louisiana during the summer of 2016 showed how extreme the consequences can be. Although Hurricane Matthew had lost power when it hit North Carolina on October 9, 2016, the floods that followed were more severe than any in the previous

half century. What will happen when a hurricane is supercharged by worsened climate conditions? On the Gulf Coast, the destruction of industrial plants would make the floodwaters toxic.

- The Pacific Ocean, at least along the coasts of Washington and Oregon, is naturally acidic. Thus, it is particularly vulnerable to a top-up of climate change–induced acidification (see earlier discussion in this chapter).
- The largest forests in the country are already under pressure from pests and giant fires, both of which are whipped up by increasing temperatures. With temperatures increasing even more, those forests will be wiped out.
- Due to its geographic position, Alaska, where melting glaciers cause giant landslides, is on the forefront of climate change. After a visit there in September 2015, President Obama made the following observations: "I saw it myself in our northernmost state of Alaska, where the sea is swallowing villages and eating away at shorelines, where the permafrost thaws and the tundra is burning. Where glaciers are melting at a pace unprecedented in modern times. It's a preview of our future if the climate changes faster than our efforts to address it."[11]

However incomplete, this review shows that complacency would be misplaced, as the U.S. military is well aware: "Climate change will affect the Department of Defense's ability to defend the Nation and poses immediate risks to U.S. national security. The Department is responding to climate change in two ways: adaptation, or efforts to plan for the changes that are occurring or expected to occur; mitigation, or efforts that reduce greenhouse gas emissions" (Department of Defense 2014). But how to relocate naval bases, particularly the world's largest, which happens to be in Norfolk, Virginia?

CHARGING UP!

These graphs show the accelerating degradation of key climate change parameters. They, of course, don't show the future occurrence of tipping points—that is, abrupt changes that are inevitable under

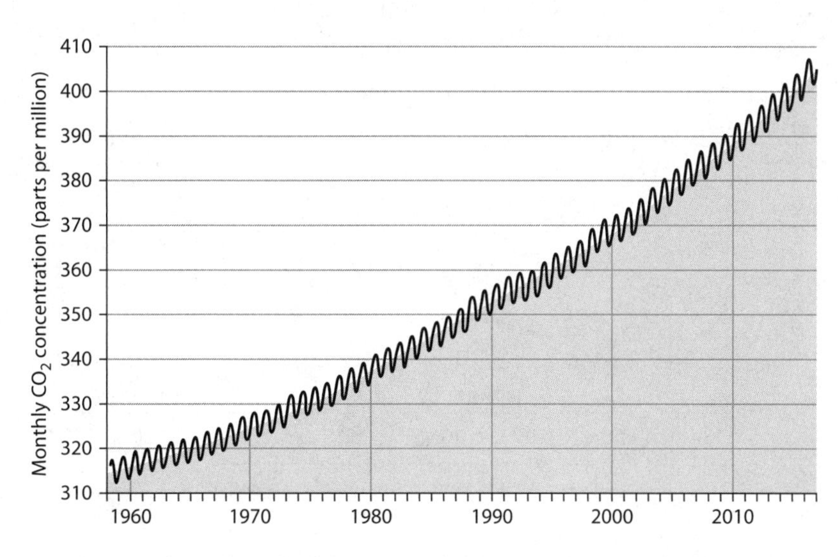

FIGURE 4.1 Carbon dioxide concentration at Mauna Loa Observatory

Source: Scripps Institution of Oceanography. http://scrippsco2.ucsd.edu

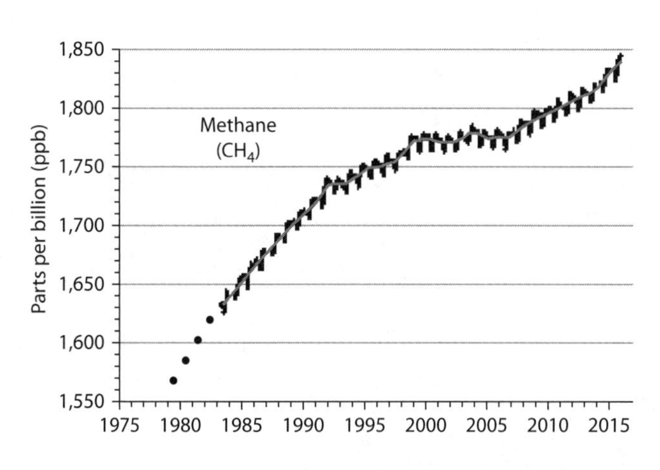

FIGURE 4.2 Methane parts per billion concentration

Source: NOAA annual greenhouse gas index. http://www.esrl.noaa.gov/gmd/aggi/aggi.html

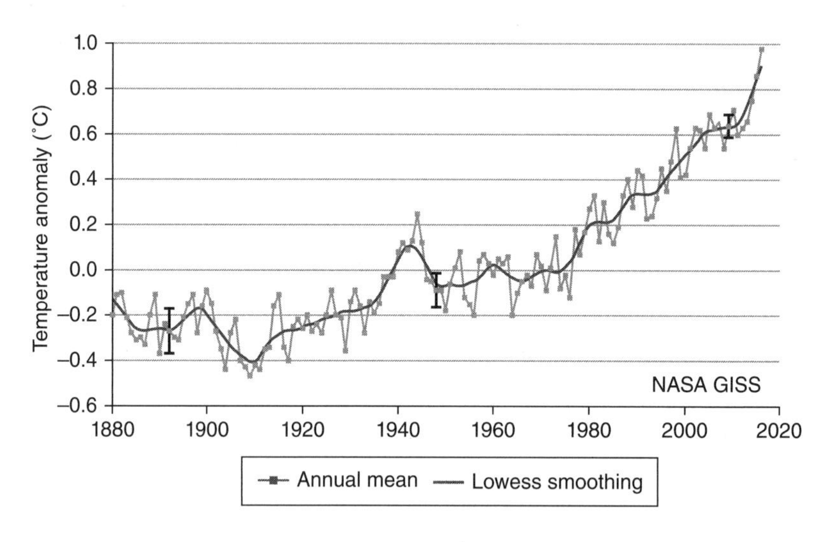

FIGURE 4.3 Average earth temperature: Global mean estimates based on land and ocean data

Source: Goddard Institute for Space Studies (GISS) surface temperature analysis, October 14, 2016. http://data.giss.nasa.gov/gis.temp/

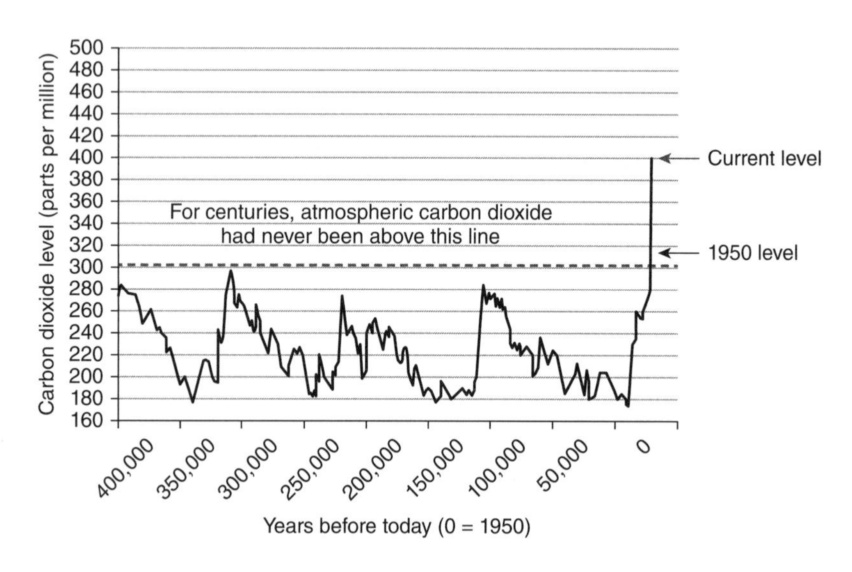

FIGURE 4.4 Carbon dioxide level (parts per million) across 400,000 years

Source: NASA Global climate change: Vital signs of the planet-evidence. http://climate.nasa.gov/evidence

the current trends. Although they cannot be assigned accurate dates of occurrence, a catalogue (Drijfhout et al. 2015) has been established on the basis of the broad selection of models used by IPCC Groups 1 and 2. Some are long-term threats. The following are examples of events expected to occur during the present century:

- Total collapse of convection in the North Atlantic, resulting in modifications in ocean circulation, possibly at long distances
- Abrupt snow melt on the Tibetan Plateau, with modifications in regimes of large rivers in China, India, Pakistan, and Bangladesh
- Almost uniform aridity around the Mediterranean Sea, the southwestern United States, and Central Asia
- The savanna-ization of the Amazon Basin

Burning fossil fuels is the main culprit. Industrialized agriculture, including animal raising, is a distant, albeit solid, second. It is responsible for slightly less than 25 percent of total GHG emissions, with major contributions to the increasing concentrations of CH_4 and N_2O in the atmosphere.

During his September 2016 visit to Hawaii, President Obama recalled meetings at the White House: “My top science adviser, John Holdren, periodically will issue some chart or report or graph in the morning meetings, and they are terrifying” (Davis, Landler, and Davenport 2016). That doesn’t sound exaggerated.

5

ENLISTING THE SCIENTIFIC METHOD

IT IS obvious that devising, organizing, and implementing scientific and technological innovation is trickier today than it was when modern science emerged. However, the methodological core of scientific and technical innovation is a lasting legacy, from Galileo Galilei to Isaac Newton to many others who followed, including Antoine Lavoisier, the pioneer of chemistry, and Louis Pasteur, a pioneer for life sciences. The legacy entails organized and systematic confrontation of interpretative ideas (called *theory*, which nonscientists too often dismiss as esoteric and useless—or worse) with aspects of the natural world that these ideas are meant to make intelligible. As mathematician Henri Poincaré put it, you don't build a house without bricks, but a heap of bricks is not a house.

Perched on the shoulders of Galileo, Newton showed the formidable reach of the scientific method when he formulated the theory of universal gravitation. This theory is not only broad but also specific in formulating a great diversity of predictions—pointing to all sorts of experimental verifications and applications.

That the roads to specific applications often go through remote and surprising territories is a lesson that has certainly not lost its relevance today, especially as far as promoting sustainability is concerned. Who, for instance, would have guessed that astrophysics—or, more precisely, astrochemistry—would be the gateway to nanoscience and nanotechnologies and thus to new materials that, inter alia, save natural resources and energy, while also generating serious dangers? The exceptional durability of the scientific method—with those recent adjunctions that use computerized models and simulations and that deal with genuine uncertainty (see chapter 7)—and its relevance in guiding a shift toward more sustainable development justify why its emergence and recent developments are laid out here, not as a piece in the history of science but as a tool to meet present and future requirements.

It is impossible to understand how science might be put to use if we don't first understand how it emerges and functions. A quick look at how the founding fathers of science were able to progress is invaluable in this respect.

Galileo's life and lasting scientific and technical contributions are of special interest. As a creative scientist in both theoretical and applied fields, an entrepreneur, a communicator, and finally a victim, he embodied the roles that are expected of present-day scientists—or that are inflicted upon them. As Alison Davis-Blake (2015), former dean of University of Michigan's Ross School of Business, put it: "Galileo taught us the importance of evidence-based management, demonstrated by his rejection of an accepted theory and use of his telescope to gather evidence." There is no better approach to the scientific method and its offspring than to briefly assess Galileo's achievements.

It is essential to then understand how, on the basis of the scientific method, scientists, inventors, and entrepreneurs have sustained a tightly knit process of mutual inspiration. The results obtained by James Watt in improving the performance of the steam engine, and the limits he faced when he wrongly thought that the way to gain power in the vessel was to increase the pressure of the steam, led Nicolas Léonard Sadi Carnot, a young officer in the French Army—who found military life boring but who was interested in thinking further about the laws of physics that he had been taught at the École Polytechnique—to the formulation of the laws of thermodynamics. One consequence of his second law of thermodynamics is that the power of a steam engine increases not with the pressure but with the difference in temperature between the cold and hot sources within the vessel. That finding, in turn, gave George Stephenson the clue to reduce the size of the steam engine without losing power so that it could be put on rails.

In the same vein, though much later, the haphazard success of inducing catalytic reactions in industrial chemistry stimulated the process of harnessing quantum chemistry, making possible a deep and operational understanding of why and how catalysts function. Catalytic reactions are sustainability friendly insofar as they require far less energy than the high-temperature reactions they replace. This is just one example of a systemic mode of interaction that may be harnessed for the benefit of a more sustainable mode of development.

As economists often say, however, there is no free lunch, which here means that those most useful scientific and technical advances also have their dark sides. Within the heritage of quantum mechanics and genetics are weapons of mass destruction, as well as a number of lesser, albeit still significant, inconveniences. Nanoparticles, for example, make materials with some very useful properties; however, when loose, they might be a greater health hazard than asbestos. Introducing useful properties in plants is one thing; genetically enhancing human beings is quite another matter. The capacity of democratic control, including proper regulatory procedures and institutions (also in the international arena), to cope with these issues will be crucial. As Sheila Jasanoff (2005) put it: "It is no longer possible to deal with such staple concepts of democratic theory as citizenship or deliberation or accountability without delving into their interaction with the dynamics of knowledge creation and use" (6).

Interfering with knowledge creation might not be possible: laws of nature are no matter for voting, and it is almost impossible to plan and control the ways in which they are discovered, as this chapter shows again and again. However, properly controlling the uses of such laws is compulsory if society is to fulfill the ambition of properly orienting the trajectory of its own development. Doing so will be extremely difficult; presently, the only convincing available model is the way in which civil nuclear energy is regulated in a few countries that take the matter most seriously (see chapters 3 and 7).

Finally, there is an even deeper concern. The physical microworld, which quantum mechanics has deciphered, and the biological microworld, which is the subject of molecular biology, have in common a fundamental and disturbing property: they are astonishingly malleable. Manipulating nanostructures or biomolecules proves easy but also leads to all sorts of transformations, some of which contribute to sustainable development, while others are deeply disruptive.

Is nature a kind of machine that, by finely targeted interventions, can be deconstructed, repaired, modified, and reconstructed at will? Or is it also an organism that is better approached in a different scientific tradition? This dichotomy is discussed at the end of the chapter.

ORIGINS AND DEPLOYMENT OF THE SCIENTIFIC METHOD: BACK AND FORTH BETWEEN SCIENCE AND TECHNOLOGY

ARTICULATING THEORY AND EXPERIMENTATION

Galileo was born in 1564—the same year as William Shakespeare—in Pisa, a city that was the economic hub of the Grand Duchy of Tuscany. His father was a well-regarded musician and mathematician who devoted much time and attention to his son's education.

At the age of seventeen, Galileo was well prepared to enter the University of Pisa as a student, but once there, he was in for a serious disappointment. While his father had developed in him a broad intellectual curiosity, he discovered at university that you basically received the same mixture of Roman Catholic theology and Aristotelian philosophy, regardless of the course of study you chose. It was all dogma, with neither experimentation nor even observation. Galileo intended, on the contrary, to "read in the book that Nature keeps open for those who use their eyes and their brain," as he put it later. His Greek model was not Aristotle, the profuse philosopher; rather, it was Archimedes, the rigorous and imaginative physicist and engineer.

More than the famous leaning tower (the campanile of the cathedral),[1] the duomo of the cathedral in Pisa was where Galileo developed his approach to the understanding of natural phenomena.

Suspended with a long rope from the top of the dome was a lamp that would oscillate after a gust of wind (the lamp is still there). Galileo's attention was drawn not by the obvious fact that the amplitude of oscillations decreased as time passed after the cause had ceased to be felt (it was merely a passing burst of wind), but by the less obvious fact that the period of the oscillations seemed to remain constant. Galileo then started a scientific inquiry that Laura Fermi, a historian of science and the wife of Enrico Fermi (one of the giants of twentieth-century physics) aptly described:

> He gathered information through observing the phenomenon (the swinging lamp) and tried to grasp its essential features. Then he went further. He formulated a hypothesis: perhaps all oscillations, of great and of small amplitude, lasted the same time. After a first rough check of this hypothesis (comparing the times of swings with his own pulse beats), he devised an experiment. The experiment reproduced the essential elements of the phenomenon albeit under controlled and simplified conditions. The complicated shape of the lamp might have influenced its motion, so he replaced it with bodies of the simplest shape, a round ball, hanging from strings whose length could be measured exactly. Only after obtaining the same result over and over again under these controlled conditions would Galileo be satisfied that the hypothesis was valid. (Fermi and Bernardini 2003, 20)

For ten years, Galileo worked in Pisa in relative marginality. Articulating repeated experimentation with progressively improved formulation, he established the first law of motion and partly the second one, of which Isaac Newton would later give a complete formulation. Then came a piece of very good news: at the age of twenty-eight, Galileo was offered a *cathedra* (a full professorship) at the University of Padua.

THE "PADUA VALLEY"

At that time, the University of Padua was the official university of the Republic of Venice. Although the doge and his deputies didn't want to have students in the middle of Venice itself, they were interested in having a renowned university. The professors at Padua enjoyed much more intellectual independence than in most other Italian universities (with Bologna as the other exception). The fact that the Republic was a fiercely independent power, often at war with the Papal States, contributed to the independence granted to the university.

Academic quality was a tradition at the university, as illustrated by a gallery in the great hall, which displays portraits of such alumni as the Polish astronomer Nicolaus Copernicus, the Belgian doctor in medicine Andreas Vesalius (pioneer of anatomy), the British doctor in medicine William Harvey (who showed how the blood circulates in the body), and the Italian poet Torquato Tasso. They testify to a long tradition of free creativity that, along with the university's relaxed atmosphere, quiet location, relatively small size, and comfortable endowment, suggest a parallel between sixteenth-century University of Padua and present-day Stanford University. Galileo's activities would, at a very modest scale of course, extend the parallel to Silicon Valley, producing the first modern example of systematic interplay between fundamental and applied science.

Copernicus and Galileo never physically met, as Copernicus was a student at Padua a century before Galileo started his lectures. But Galileo had a copy of Copernicus's *De Revolutionibus Orbium Coelestium*, in which the hypothesis is made that the earth orbits around the sun and not the other way round. Copernicus was so afraid of an inevitable condemnation by the Church that he delayed the publication of his book. He did not get the first published copy until he was on his deathbed.

Galileo did not have such fears. He was confident in himself and in the persuasive power of scientific evidence. In addition, unlike his predecessor, Copernicus, and his successor, Newton, he was an excellent communicator. However, he wanted to be absolutely sure that Copernicus's hypothesis was vindicated by observed facts. The only way to do so was to look at the sky in a way that nobody had done before him.

Galileo did not invent the telescope, but he did invent the first truly efficient telescope, equipped with two opposite lenses—one concave and the other convex. This combination increased the magnifying power tenfold. Being an astute communicator, he leaked some information about his telescope, and that information reached the doge, who summoned Galileo for a demonstration. Together they went up the bell tower above the Piazza San Marco, in the heart of Venice. The doge was stunned: he could see ships two hours' sailing time away as if they were entering the lagoon. He quickly grasped the military and commercial potential of the instrument. Soon after, Galileo was again summoned by the doge, this time to attend a grand ceremony in his honor at the Doge's Palace, where he was granted tenure for life, with a salary unmatched in any Italian university.

His new role contributed to broadening his fame and to starting a profitable business. The Republic placed no embargo on the telescope, so Galileo was able to create a sort of cottage industry of optical and mechanical devices with the best artisans in Padua—a small Silicon Valley of the time that became known as far as China. The devices not only sold very well and helped finance Galileo's experiments but they were also of direct use for him, especially the telescope, which he turned toward the sky.

The discoveries he made were of an incredible diversity: he discovered mountains and seas on the moon, the four satellites of Jupiter (always the astute communicator, he called them the Medici planets in honor of the Grand Duke of Tuscany), sunspots, and most fatefully the phases of Venus. Most fatefully, indeed: the succession of the phases that he observed proved incompatible with an earth-centric model. As such, Aristotle, Ptolemy, and the Church were categorically proven wrong, as just one contradiction definitely invalidates a hypothesis. The phases were compatible with a heliocentric model. Thus, Copernicus was provisionally right—provisionally because, Galileo insisted, a hypothesis that is vindicated by one set of observations or experiments may subsequently be invalidated by new observations or experiments. This last point is a fundamental tenet of the scientific method. A clash with Rome was looming—the scientific method versus the dogmatic mantle of vested interests.

Galileo published his observations and conclusions in *Sidereus Nuncius*, which was first circulated in 1610. He was, of course, aware that they ran against the traditional position of the Church, but he believed in the persuasive power of scientific evidence, as he wrote to the grand duchess Christina in 1615:

> I do not feel obliged to believe that the same God who has endowed us with senses, reason, and intellect has intended to forgo their use and by some other means to give us knowledge which we can attain by them. He would not require us to deny sense and reason in physical matters which are set before our eyes and minds by direct experience or necessary demonstrations. (Drake 1957)

In fact, Galileo misjudged the main reasons the Church was desperate to uphold the orthodoxy that the earth stood at the center of the universe. The Church knew that it was the image of the central place of man in God's creation and of the central place of the Church in human affairs. Thus, it could not accept a scientific approach that might prevent the Church from deciding the order of things.

At first, it seemed that Galileo's optimism was warranted. During a visit to Rome, he was greeted as a hero by the populace and welcomed by an influential faction of the Church hierarchy. But the Inquisition was preparing a counterattack and, in 1616, demanded that he end his support of Copernicus's hypothesis. Treating the matter as if it were a scientific controversy, Galileo responded in 1633 by publishing *Dialogue About Two Major World Systems*, which was all the more unacceptable for the Inquisition because it was highly readable. At the age of sixty-nine, Galileo was charged with heresy and summoned to the Tribunal of Inquisition. As a faithful Catholic, he abided by the injunction. Pope Urban VIII was no longer in a mood of tolerating Galileo's innovations; he was on the same line as the Inquisition. Thus, Galileo basically had the choice between abjuring voluntarily or abjuring under torture. He was then condemned to house arrest in Florence for the rest of his long life.

While in Florence, he received a visit from a young English poet who was soon to become famous with his *Paradise Lost*. John Milton recalled that visit in *Areopagitica*, circulated during a successful campaign in England against a bill before the English Parliament that would have allowed censorship of scientific results: "There it was that I found and visited the famous Galileo grown old a prisoner to the Inquisition, for thinking in Astronomy otherwise than the Franciscan and Dominican licensers thought" (Milton 1644).

A SYSTEM OF THE UNIVERSE

Isaac Newton was born on Christmas Day in 1642, the year Galileo died. When he graduated from university twenty-four years later, in 1666, Newton had read and assimilated all of the books that could be of any help for his future work (as he said, "If I have seen far, it is by standing on the shoulders of giants"; see chapter 8). He had even started to lay down a new branch of mathematics called *differential calculus*, which was to be the natural language for formulating the laws of universal gravitation. Just as he was poised to become a don at his Cambridge College, Trinity, England experienced an *annus horribilis*. Not only was 1666 the year of the Great Fire in London; it also saw the resurgence of the Black Death. Cambridge University was closed down, and Newton took refuge in Woolsthorpe, the village in Lincolnshire where he was born. For two years, he was deprived of what scientists tend to consider an essential ingredient to creative work: a scholarly atmosphere. Yet, for those two years, as Newton worked in complete solitude, he turned what would have been a major hindrance to most people into a factor of success.

Thus, two years after graduating, at the age of twenty-six, Newton had built one of the greatest scientific monuments in human history—universal gravitation—along with the relevant mathematical language. He had given the scientific method its full stature. The range of the theory is stunning: it encompasses the revolution of planets, the return of comets, the tides of the oceans, the movements of all objects on earth (from the fall of the mythic apple in Trinity Garden to ballistics—the military is never far away when science innovates). His work led to a wealth of potential experiments and inventions. The theory has a unique quality of universalism, which inspired Alexander Pope in writing Newton's epitaph (March 21, 1727): "Nature and Nature's Laws lay hid in Night: God said, 'Let Newton be!' and all was light."

Newton himself proved better aware of what science actually is, when writing to a friend:

> I do not know what I may appear to the world. But to myself I seem to have been only like a boy, playing on a sea-shore, and diverting myself in now and then finding a smoother pebble, or a prettier shell than ordinary, whilst the great ocean of truth lay all undiscovered before me. (quoted in Brewster 1855)

What a collection of shells he found! Yet other remarkable shells were lying elsewhere, waiting to be discovered through the scientific method: electricity, chemical bonds and reactions, evolution of species, special and general relativity, quantum mechanics (which led to exclude Newton's theory from the microworld), genetics, and more. It is a richly productive ocean from which it is a paramount task of our time to catch anything that might contribute to making development more sustainable, while also containing undesirable, and sometimes highly dangerous, potentialities.

What Galileo, Newton, and other scientists provided is what economist and historian Joel Mokyr (2004) calls the epistemic base of technological developments:

> When no one knows why things work, potential innovators do not know what will not work and will waste valuable resources in fruitless searches for things that cannot be made, such as perpetual-motion machines or gold from base metals. The range of possibilities for experimentation that needs to be searched over is far larger if the searcher knows nothing about the natural principles at work. To paraphrase Pasteur's famous aphorism . . . , fortune may sometimes favor unprepared minds, but only for a short while. It is in this respect that the width of the epistemic base makes the big difference. (Mokyr 2004, 31)

And the relationship is two-way, Mokyr added. Alluding to the articulation between the development of the steam engine and the formulation of the laws of thermodynamics, he said: "There was feedback from techniques to propositional knowledge" (Mokyr 2004, 54).

Let us now turn to more recent stories which in the same spirit might, if properly used, be essential stones in the foundations of a more sustainable world.

CONTRIBUTIONS OF QUANTUM SCIENCE TO SUSTAINABILITY—AND TO COLLECTIVE DISRUPTION

According to Mokyr, who refers to a paper in *Scientific American*, 30 percent of the U.S. gross national product proceeds from applications of quantum mechanics—that is, the physics of the microworld. That future did not seem to be in the cards when this seemingly strange and abstract approach to the physical world was developed in the 1920s and 1930s. It is not much of an exaggeration to say that, apart from the forty or so scientists who were directly involved in this adventure, nobody was aware that a scientific revolution was underway—not to mention the idea that far-reaching industrial and societal transformations would ensue.

Although the scientists involved enjoyed lively controversies among themselves, they were confined to scientific and philosophical interpretations of the new physics. The most famous and most relevant were Niels Bohr and Werner Heisenberg on one side and Albert Einstein and Louis de Broglie on the other; their dispute was about the significance of Heisenberg's uncertainty principle, which bars predictability at the individual level. According to this principle, it is impossible to simultaneously know with complete precision all the parameters attached to a quantum object (such as an electron or a photon), including its position and velocity (see chapter 7). Nevertheless, quantum mechanics provides objective probability distributions that make all the phenomena it encompasses fully predictable statistically. That is indeed the way we perceive them when using electronic devices or lasers, for example. These devices run not on isolated electrons or photons but on enormous flows of these particles, flows for which statistical regularities provide working stability. Never has the uncertainty principle popped up through a computer screen, and yet we know that the computer screen works because of these particles.

At Bell Labs at the outset of World War II, William Shockley, John Bardeen, and Walter Brattain put electrons firmly to work. They extended the reach of quantum mechanics to solid-state physics, starting with crystals, which are solids with stable regular structures. They also pursued practical goals that may be considered as sustainability goals (though that term was unknown to them): saving energy and improving reliability of signal transmission along the telephone lines of the Bell system. From quantum mechanics, it was known that, if a tiny electric current (a flow of electrons) is applied at one end of a crystal (in this case, germanium and silicon), many more electrons are pulled out from the envelopes of the atoms within the crystal, so that a highly amplified current comes out at the opposite end of the crystal. This phenomenon would have been impossible to explain within Newtonian physics; it is thus a good illustration of the inadequacy of Newtonian physics to the quantum world. Their device effectively worked at the end of 1947, and the three physicists called their invention a *transistor*.[2]

The transistor did indeed prove efficient in sustaining signals along the Bell lines, but it also had many more uses, as became clear later. It was to be the cornerstone of the so-called information and communication society. Electronic communication is not perfect from a sustainability point of view, but it is perfectible (reducing the waste heat from all kinds of machines involved—in particular, the big servers—is an example of improvement in this respect; reducing the amount of futile communication would be another). In addition, it is often preferable to physical moves. Moreover, along the journey from the first transistor to the sophisticated devices and networks we have today, William Shockley, Robert Noyce, Steve Jobs, and others invented a new mode of promoting and organizing innovation—call it the Silicon Valley model. Some of their successors now explicitly try to orient this model toward innovations with a high sustainability potential, such as the conversion of light into electricity and fossil fuel–free modes of transportation.

What Shockley had initiated at Bell Labs with the electron, Charles Townes did a few years later at Columbia University's department of physics with the photon. Townes invented the precursor of the laser (light amplification by stimulated emission of radiation) because he wanted to better see the architecture of complex molecules. The coherence of the beam produced (all the waves in a beam coming out of a laser have the same length) makes the laser a high-resolution spectrometer. His initial objective undoubtedly made for an interesting application, but it is a far cry from the present ubiquity of the laser in all sorts of applications and at all levels of power that Townes (1999) describes with a measure of surprise.

Electrons have been put to work in ways that, grosso modo, are not inimical to sustainability; so too have photons. It now appears that a raft of micro-objects previously unknown or overlooked might also play various significant roles: it is the time of the nanorevolution, with promising prospects for sustainability, albeit with some rather dark sides as well.

At the University of Sussex, Harold Kroto discovered unknown chainlike carbon molecules in interstellar gas clouds. Would it be possible to vaporize carbon on earth to find out how exactly these molecules are made? To try to answer this question, Kroto headed to Rice University in Houston, Texas, where Robert Curl and Richard Smalley had a very powerful laser that could vaporize almost any known material. In a series of experiments performed in the autumn of 1985, the three of them simulated physicochemical reactions that take place in interstellar clouds by applying the power of their laser to the most common form of carbon—graphite. The results were stunning. Hitherto unobserved, and altogether remarkable, carbon molecules were detected—in particular, nanotubes and the now emblematic buckyball. The buckyball is an extremely stable carbon molecule in which sixty atoms are arranged in three-dimensional space as the nodes on the geodesic domes that architect Buckminster Fuller had invented and built in many places. The pavilion for the United States at the Montreal Universal Exhibition, now the city's Biosphere, is a well-known example.

Kroto, Curt, and Smalley discovered that the molecules they were producing by vaporization of carbon (or later of other simple substances) display astonishing mechanical, electrical, and chemical properties. As far as mechanical strength is concerned, the nanotubes were up to two orders of magnitude greater than for steel.

Electrical properties match those of the most electrically efficient materials. The high level of chemical activity is linked to the high surface-to-volume ratio that the small size entails. In short, one might say that ordinary materials, when embodied at very small scales, display remarkable properties that cannot be anticipated at larger scales.

Their work has been the launching pad for the explosive development of nanotechnologies, with significant sustainability benefits and potentially serious hazards. Nanoparticles are used to improve—in some cases, dramatically—the performances of various materials and devices: building materials (for example, panes of glass blocking infrared radiation due to the introduction into their structure of invisible nanoparticles), solar panels with enhanced conversion rates, flat-screen electronic panels displaying clearer images, and more efficient batteries (see chapter 6), to name just a few. Nanoparticles are also used to perform new specific functions, such as delivering drugs precisely where they are necessary in the body. And as just one illustration of a number of cosmetic uses, titanium dioxide particles are now routinely used to protect the skin from sun damage. Here we have an example of the serious safety issues that nanoparticles raise: their small size and high chemical reactivity imply an unusual capacity to escape into the environment, to penetrate living organisms, and to form chemical compounds of unknown toxicity. The nanorevolution illustrates particularly well a chasm between science and its applications: because laws of nature cannot be put under democratic (or otherwise) control, it is imperative that applications be regulated, some of them tightly, within democratically established institutions.[3]

THE CONTRIBUTIONS OF BIOSCIENCES TO SUSTAINABILITY—AND TO COLLECTIVE DISRUPTION

If you are after a beer in Cambridge, UK, you can go to The Eagle; there look for a plaque bearing the names of Francis Crick and

James Watson and sit at the table where they discussed the advancement of their work at the nearby Cavendish Laboratory. The Crick and Watson model of the DNA double helix identifies its constituents (sugars and phosphates); shows how they alternate; and shows how the DNA molecule duplicates itself, generating the replication of the cell. It also shows how the molecule commands the production of proteins (Watson and Crick 1953).

Crick and Watson's model opened a broad avenue to investigating the mechanisms of life and to bending them, as Herbert Boyer and Stanley Cohen did when they were starting genetic engineering in 1973. In his Stanford University lab, at the beginning of the 1970s, Cohen was pursuing the idea of transferring genetic information from one species to another—for example, transferring into a bacterium the information that commands the production of insulin in the human body. In this example, when the bacterium reproduces, it copied the human DNA into its offspring, thus building a natural factory for the production of insulin. In 1972, Cohen had mastered several steps of the process, but he was still some distance from completion. Then, at a scientific conference in Hawaii, he attended a presentation by Herbert Boyer, from the University of California at San Francisco. Boyer reported that he had isolated an enzyme that could be used to cut DNA strands into predetermined segments: a crucial missing step in Cohen's effort. That evening, over a pizza, Boyer and Cohen informally decided to join forces. Less than a year later, the cornerstone of the initial developments in genetic engineering had been laid down, for better or for worse.

Boyer and Cohen obtained patents but licensed them broadly at moderate prices (even free of charge for not-for-profit research), thus furthering a quick dissemination of their method. Had they acted in a more restrictive way, a serious public policy issue would have emerged (see chapter 8).

These achievements, however recent and highly significant, pale in comparison with the even more recent and dazzling results in genome editing, which make the task precise, easy, and cheap. The new technique, known as CRISPR (clustered regularly interspaced short palindromic repeats), appeared as recently as 2012 and is already revolutionizing the field of genome editing (Barrangou 2014), a concept that from now on should be interpreted in a literal sense.

Before 2012, it was known that bacteria are able to fend off viruses by enlisting specific enzymes (which are now collectively called Cas, for CRISPR associates) that cut DNA sequences within the viruses at appropriate loci. In 2012, Emmanuelle Charpentier and Jennifer Doudna showed that the cleaving power of some Cas enzymes can also be put to work in all sorts of plants and animals—human beings included—in the following way:

1. A DNA sequence in a living organism is targeted for being cleaved.
2. An appropriate Cas is guided toward the target by an RNA molecule containing a sequence that matches the target sequence.
3. When the target is reached, it is cleaved by the Cas at the chosen locus.
4. A process of repair then starts naturally; at this stage, genome editing is performed.

As indicated in Charpentier and Doudna (2013, 20): "When the breaks are repaired by standard cellular repair mechanisms, the sequence at the repair site can be modified and new genetic information inserted." It is thus possible to switch, add, or remove elements such that an encoded protein is changed, or the previous role of the targeted sequence altogether disappears, or a new role is assumed. Let's consider two specific cases:

- *Getting rid of powdery mildew in wheat*: Powdery mildew is a fungal disease that causes large white spots on the leaves or stem of a plant, interfering with photosynthesis. It may spread quickly where humidity is high and temperatures moderate, and its development is boosted by the intensive use of nitrogen fertilizers (see chapter 2). It significantly reduces yields—reductions of up to 45 percent have been observed in the midwestern United States. Caixia Gao and Jin-Long Quiu of the Chinese Academy of Sciences have used CRISPR-Cas to delete from a wheat strain the genes encoding proteins that repress defenses against powdery mildew (Gao and Quiu 2014).

- *Mice with gene mutations that cause cataracts*: Yuxuan Wu and his team at the Chinese Academy of Sciences have cured such mice using CRIPSR-Cas corrections of the genes involved (Wu et al. 2013).

Wei-Feng Zhang of the Massachusetts Institute of Technology, whom *Nature* has dubbed "DNA's master editor" (2013). He intends to use CRISPR to treat neuropsychiatric conditions, such as Huntington's disease and schizophrenia, by repairing the involved genes. He says he feels "limited only by what I can imagine is possible"—an idea that is both exhilarating and terrifying.

Chinese labs are active in the field, as are U.S. labs and start-ups. From Massachusetts to California, such companies are mushrooming: Editas Medicine, Caribou Biosciences, CRISPR Therapeutics, Intellia Therapeutics, and so on. According to advisers in investments, these companies are "hot": they burn a lot of money, get even more from fascinated investors, and are not keen for regulation.

MACHINE OR ORGANISM?

Nature as an infinitely malleable machine is the ultimate embodiment of the Galileo-Newton revolution as interpreted by Carolyn Merchant (1980, 290): "The mechanistic view of Nature, developed by the seventeenth century natural philosophers . . . assumes that nature can be divided into parts and that the parts can be rearranged to create other species of being." This is indeed an aspect of reality, though not the whole of it. Nature also manifests itself as an organism in which "parts take their meaning from the whole, each particular part being defined by and dependent on the total context" (Merchant 1980, 293). The relevant scientific approach here is ecology (see chapter 6).

Systematically viewing nature merely as a machine, never to get exhausted, has a perverse effect: it aligns science with the forces busy at plundering nature. Then nature is “bound into service,” in Francis Bacon’s words, and science is compromised with

> those men who
> Ransack’d the Center, and with impious hands
> Rifl’d the bowels of their mother Earth
> For treasures better hid.
>
> John Milton, *Paradise Lost*,
> Book 1, lines 686–688

6

SUSTAINABILITY AT THE INTERSECTION OF SCIENCE AND NATURE

MANY OF the problems raised in chapters 1 through 4 derive their answers (some of them partial but still useful) from associating scientific results with nature's models. This chapter addresses answers to such questions.

For questions about biodiversity, available answers are fragmented and often lack teeth, though there are bright spots. For example, under favorable conditions, rain forests can resist aggressions or recuperate from them better than originally expected. For the oceans, properly managed zones, where the prohibition of damaging activities is clearly stated and consistently enforced, are also susceptible to impressive recoveries.[1] Also remarkable is the power of specific ecosystems, such as mangroves, marshes, and oyster and coral reefs, to protect coasts against erosion and the effects of extreme meteorological events, as well as to match sea-level rises. Genetically modified organisms are additions—albeit often unwelcome ones—to biodiversity. Some such organisms indeed offer benign benefits for the natural environment—for instance, rice strains made more resistant to a lack or excess of water or salinity.

For water and soil, articulating science and nature provides radical improvements. Significant water savings have been made possible through properly conducted irrigation and elaborate recycling techniques. Soil sustainability is conditional on the substitution of integrated biological approaches for chemical treatments.

The section of this chapter dealing with energy focuses on how to make the transition away from fossil fuels as efficiently and quickly as possible. Three approaches are considered: maximizing energy savings, promoting new and more efficient concepts of solar cells, and balancing the intermittency of renewable sources of energy with much improved storage devices.

The chapter closes with a discussion of natural and artificial ways of capturing and storing CO_2, the importance of which is increasingly recognized. According to conclusions of a report produced by the National Research Council (2015) for the U.S. National Academies: "It is clear that atmospheric CO_2 removal is valuable, especially given the current likelihood that total carbon emissions will exceed the threshold experts believe will produce irreversible environmental effects" (111). The German Institute for International and Security Affairs (or SWP, Stiftung Wissenschaft und Politik) concurs: "Without negative emissions, the 1.5°C target would be out of reach, and the 2°C could only be met at far greater cost, if at all" (Geden and Schäfer 2016, 4).[2] Myles Allen—professor of geosystem science at the University of Oxford and co-author of a landmark article on the overall limit of CO_2 emissions compatible with a mean temperature increase kept below 2°C (Meinshausen et al. 2009; see also chapter 3)—went so far as to urge that for every ton of CO_2 emitted, one ton be captured and stored, with the emitter covering the expense (Allen 2015). Negative emissions are no doubt crucial, as also stressed in the *Fifth Assessment Report* (IPCC 2014), and yet political and economic actors systematically overlook the issue.

BIODIVERSITY: A SOURCE OF INSPIRATION AND PARTNERSHIP

PATCHES OF RESILIENCE IN RAIN FORESTS

In 1997, when the government of Guatemala started granting indigenous communities concessions spanning twenty-five years for the management of rain forests, the decision stirred controversies. Such communities were, in general, not seen as being up to the task, compared with specialized public bodies and experienced enterprises. However, strict conservation laws don't always prevent neglect in public management and surveillance. Likewise, however experienced, many enterprises often have too narrow a scope.

Indeed, fifteen years later, the skeptics were proved wrong. An in-depth audit of the controversial concessions in the Maya Biosphere Reserve (MBR), which extended around the remains of ancient Tikal, concludes:

> The core finding of this study is that the timber harvesting in the MBR is sustainable and in fact represents state-of-the art best practice globally for species-level management in tropical forests. At present levels of harvesting, populations of commercially important timber species are expected to recover initial densities and volumes during cutting cycles between successive harvests, on average. Such a finding, backed up by scientifically rigorous, field-based empirical data, sets the MBR apart from most other commercial forestry operations in the tropics. (Grogan et al. 2015, 4)

Most indigenous communities in tropical forests have only weak rights and precarious tenures over where they live; hence, they have neither the incentive nor the capacity to manage the forest sustainably or to resist the pressures of outside forces interested in mining the forest rather than ensuring its reproduction (Gray et al. 2015). Secure tenures, however, provide communities with the confidence and power needed to effectively manage their lands. In this respect, even twenty-five years of concessions are too short, as this amount of time does not match the cutting cycles of the dominant timber species, though this hasn't deterred sustainable management and conservation in the MBR.

State-of-the-art forestry, as practiced in the MBR concessions, implies close attention to the compatibility between natural rhythms and harvesting practices. Cutting cycles are planned on a long-term horizon such that the composition of the forest, on average, is time invariant. This idea precludes the mining of commercially most-valuable trees—in the MBR, mahogany and Spanish cedar—and the neglect of other species, such as the manchiche, for which new uses and customers have actively been sought and found (Bronx Zoo, New York Aquarium, Leroy Merlin, which is a large European network of specialized outlets). As a general rule, harvesting is conducted in ways that are in harmony with species' biology—especially in terms of regeneration requirements.

The community of the Amazonian Suruí Indians is not hindered by concessions limited in time. In 1988, a revision of the Brazilian constitution recognized the rights of Indians on their original lands, including the right to exclude intruders. However, the path from complete isolation in the forest to this recognition has been long and is paved with horrendous episodes.

Isolation of tribes in the Amazon ceased in the 1960s, with the construction of the first road through the rain forest—the Trans-Amazonian Highway. The first contacts between the Suruí and the outside world were devastating: transmissible diseases against which they were defenseless and bloody confrontations with would-be settlers, who were lured from all over Brazil by fraudulent property titles on Amazonian land, decimated the population from two thousand to a few hundred. Under similar circumstances, many other Indian communities in the Amazon were wiped out. With characteristic energy and under an exceptionally visionary leadership, however, the Suruí bounced back. In 1988, when their exclusive rights on their lands were confirmed, they proved able to make a living from them in sustainable ways:

- They harvest timber and nontimber products according to rules similar to those implemented in the MBR. Their tenure security made the Suruí confident enough to devise and implement a fifty-year plan for harvesting and regenerating the forest and to reforest those parts of their lands that, during the extremely difficult pre-1988 period, had been devastated by logging and criminal fires.
- They share the cost of storing carbon with the outside world. The Suruí started to reforest on their own, but that effort could not be pushed very far without financial resources from the outside.

Chief Almir Suruí—who manages to combine a deep attachment to the ancestral knowledge and traditions of his community with a university education (in biology) and an interest in advanced technologies and who is highly regarded both by the members of his community and by many partners in Brazil and abroad—considers it legitimate to seek contributions from outside the Suruí community. Having been granted verified access to the markets for carbon offsets, the Suruí community agreed, in September 2013, to an initial transaction amounting to 120,000 tons of carbon offsets bought by the Brazilian company Natura Cosméticos, which is Latin America's largest cosmetic company and which aims at complete carbon neutrality.

- They protect the forest from further degradation. Illegal logging and criminal forest fires are ongoing threats that are essential to detect and repress as quickly as possible, before damages become too large. Detection is the task of the Suruí themselves. They use advanced positioning systems under a specific application of Google Earth Outreach. As soon as suspicious activities are detected, federal police agents are sent exactly to the right place.

The whole world is benefiting from the conservation and regeneration of the Amazonian forest. Seen from the sky, the Suruí lands appear as a green oasis surrounded by parched and ocher earth that was once forest.

Of course, criminal activities are never completely eradicated. There is even a contract on Chief Almir's head, a threat that is taken seriously not only by the Suruí but also by the federal authorities, who decided to assign bodyguards from the police special forces. Nobody has forgotten the assassination on his doorstep of Chico Mendes, the legendary defender of the forest and the poor.

COASTAL PROTECTION: NATURAL AND ARTIFICIAL, OPPOSED OR COMBINED

Mangroves and other coastal forests, marshes, and coral and oyster reefs are ecosystems that, in addition to a variety of ecological services (Gittman et al. 2016), are partners in the protection of coasts. By dissipating at least part of the energy carried by waves, they reduce the constant erosion of the coast; in rough weather, they reduce the force of the impacts of storms on natural structures and artificial protection barriers. In the Netherlands, where huge artificial structures are essential for protecting the polders that lie below sea level, engineers now lay marshes and shellfish beds in front of these structures. Mangroves and marshes slow the flow of water, which allows sediments to settle. Sedimentary accretion in mangrove forests more or less keeps pace with rises in sea level. Oyster reefs have the additional advantage of growing biologically; on the basis of observations made over a fifteen-year period on the U.S. Atlantic coast, Rodriguez et al. (2014) concluded that oyster reefs "should be able to keep up with any future accelerated rate of sea level rise."

On the basis of these scientific results, there is justification for restoring the salt marsh islands in Jamaica Bay, New York, into natural storm barriers. Architect Stephen Cassell is even more ambitious with his plan for protecting Lower Manhattan, including setting a garland of marshes around the financial district: "We weren't fully going back to nature with our plan. We thought of it more as engineered ecology. But if you look at the history of Manhattan, we have pushed nature off the island and replaced it with man-made infrastructure. What we can do is start to reintegrate things and make the city more durable" (Rieland 2013).

In 2011, when North Carolina's shores were hit by Hurricane Irene, three-quarters of the artificial seawalls and bulkheads were severely damaged and failed to protect the properties behind them, whereas marsh grasses efficiently protected the shore and naturally returned to their prehurricane condition within a year. Similar contrasted observations were seen on the shores of Chesapeake Bay after Hurricane Isabel in 2003 (Popkin 2015). The entire U.S. coastline would benefit from the same approach, as suggested by a broad empirical study by Arkema et al. (2013), who also participated in the Natural Capital Project at Stanford University. Indeed, 67 percent of the U.S. coastline still benefits, to some extent, from natural protections, the loss of which would make millions of people vulnerable to the vagaries of the sea.

Bangladesh is suffering from the effects of sea storms more than any other country, and the threats from sea level rises are daunting—particularly at and around the mouth of the Meghna River on the southwestern edge of Bangladesh. It so happens that one of the largest and lushest mangrove forests in the world, the Sundarbans, sits on an island facing the mouth of the river. The Sundarbans are "so thick that history has hardly ever found the way in" (Rushdie 1981). But history might soon find a way around the Sundarbans: there are advanced plans for building a large coal-fired power plant within 15 km of the forest. Both the pollution and the shipping traffic bringing coal from India would upset the fragile ecological balance, illustrating yet another manifestation of the curse on natural capital.

THE GREAT FLUORESCENT PROTEIN

In life sciences, learning to innovate from natural models—in this case, a micromodel—drawn from biological diversity, is often called *bioinspiration*. The road to fame of the green fluorescent protein (GFP) illustrates where learning from a natural model may lead in terms of sustainable innovation. In the late 1950s and early 1960s, Osamu Shimomura (of Nagoya University and Princeton University), a survivor of the Nagasaki atomic bombing, observed how *Aequorea victoria*, a jellyfish living off the coasts of Washington state, emits green light. He identified the protein responsible for the emission and called it GFP.

Thirty years later, Martin Chalfie (of Columbia University) hypothesized that it might be possible to have GFP expressed in other living organisms and still emit the green light. This view was met with skepticism in the profession, but Chalfie succeeded with the proverbial worm—*Caenorhabditis elegans*. As Chalfie suggested in a memorable 1994 paper in *Science*, his results opened the way for GFP to become a universal biological marker (Chalfie et al. 1994). Roger Tsien (of the University of California, San Diego) proved Chalfie right. Tsien even engineered several variants of GFP that emit a range of different colors. A revolution in medical imaging ensued, sustainable in the sense that it is sober in resource requirements and has no harmful side effects. GFP is also used to detect the presence of inappropriate components in drugs and foodstuffs.

BEWARE OF DARWIN

Working with nature as a source of inspiration or as a partner brings great benefits. Working against nature—inadvertently or out of negligence—may have disastrous consequences, as the collision of certain genetically modified plants (GM plants) with Darwinian selection testifies.

GM plants are additions to biodiversity that might have significant economic, social, and even ecological value. For instance, rice or wheat made more resistant to drought (or to an excess of water or salinity) would change the global food equation, while also saving valuable natural resources. However, it is not in this spirit that GM plants have been designed and marketed. Instead, the main reason for their use is to be found in the economics of monopoly behavior—or, more precisely in the present case, the behavior of a strongly coordinated oligopoly. It is well known that the U.S. firm Monsanto has a strong grip on the production and commercialization of GM plants, due in part to the fact that Monsanto owns crucial patents on gene transfers. These patents were first granted in the United States and then in all countries that Monsanto chose to penetrate (for more about patents on genes, see chapter 8).

Mostly concerned with their profitability, Monsanto and the other firms in the oligopoly were not interested in having gene transfer techniques used to meet the needs of developing countries, as tinkering with pest control in developed countries looked far more profitable. In an empirical investigation into the development of genetically modified seeds, economist Harhoff and colleagues (2001) concluded that firms like Monsanto and Syngenta have given priority to traits (for example, resistance to an insecticide or herbicide) linked to a product they already sell. One such example is the Monsanto flagship herbicide, Roundup Ready.

This is not to say that the GM plants brought to market do not deliver benefits, such as increasing yields or, to a certain extent, saving inputs like farmers' labor and conventional pesticides. It does, however, appear that, as time passes, benefits tend to stagnate, while detrimental side effects develop. Some significant cases are sketched here, with an aim to exhibit the underlying trends.

Since genetically modified cotton was introduced in India at the beginning of this century, more than seven million farmers have adopted Bt varieties—that is, varieties that have been modified by the introduction of genes from the bacterium *Bacillus thuringiensis.* These genes endow the cotton with toxic power against the bollworm, an insect that eats the cotton fibers from inside the plant. There have indeed been increases in yields, as well as decreases in the use of conventional pesticides, and farmers' incomes have been boosted (up to 50 percent in certain places). In several states, however, this trend didn't last beyond the 2007–2008 growing season.[3] Darwinian selection favored those bollworms that had developed resistance against Bt toxins, and other pests that are not targeted by the toxins benefited from reduced competition.

Chinese farmers have been confronted with similar problems. Those raising cotton in northern China have generally adopted a Bt variety producing a toxin that works well against their main pest (*Helicoverpa armigera*). The more that strategy proved successful, however, the more a secondary pest *(Miridae*), which had previously been checked by competition with the dominant pest, thrived because the toxin didn't work on them. As reported in *Science* (Lu et al. 2010), *Miridae* now attack en masse not only cotton crops but also orchards and vineyards all over six Chinese northern provinces.

Farmers in North America have also been made aware of the mixed blessings of GM plants. Monsanto engineered a variety of canola for resistance to its Roundup Ready herbicide. Roundup can thus be sprayed on canola fields without damaging the crop itself, a benefit for farmers of these fields. However, contrary to corn and soybeans, canola seeds easily germinate in all sorts of conditions. Hence, having been made resistant to Roundup (and some other herbicides), canola has become, outside the fields where it is grown, a rather inconvenient weed in North Dakota, Oregon, and Manitoba (Londo et al. 2010; Schafer et al. 2011).

Farmers in the Midwest on both sides of the U.S.-Canada border are facing comparable problems under an even more inconvenient form. A weed the size of a human adult—the giant ragweed (*Ambrosia trifida*)—has captured genes of resistance to a variety of herbicides. This development is all the more disturbing because the plant's light pollen spreads easily. Farmers must prune away the weeds individually, which takes time and raises the risk of allergic reactions.

What do all these situations have in common? When GM plants are engineered to alter some specific interactions between living organisms, there are two fundamental biological risks:

- Darwinian selection tends to boost populations developed from individuals that happen to get protected by gene mutations (Palumbi 2001).
- Altering one type of interaction might also inadvertently alter other interactions—for instance, controlling one type of pest might free an ecological niche that another type, possibly difficult to control, then occupies.

The industry not only refused to consider these perspectives but also refused to let anyone else consider them. An editorial, "A Seedy Practice," in the August 2009 edition of *Scientific American* made the point: "It is impossible to verify that GM crops perform as advertised. That is because agritech companies, abusing their monopoly power by unduly interpreting what their patents allow, have given themselves veto power over the work of independent researchers" (28).

With GM plants that are engineered for better adaptation to certain environmental conditions, however, the same problems do not arise (though this doesn't mean that producing efficient GM plants is a straightforward proposition). Instead, such GM plants prove valuable additions to natural biodiversity. They are also plants that the industry is in no hurry to develop (due to their lower profitability). Fortunately the patents on which the monopoly power of the industry rests have been circumvented by independent discoveries in the transfer of genes—in particular, at Cambia, a nonprofit institute based in Canberra, Australia (Broothaerts et al. 2005). No less important, these more effective processes have not been patented in the conventional way (i.e., to "restrict access"), but in the same way software developers have introduced with "open-source contracts" (i.e., to "keep access free"; see chapter 8). In addition, the fact that significant improvements have been made in more traditional hybridization techniques has contributed to opening the array of opportunities.

The consequences of these advances are far-reaching. As just one example, in Asia, tens of thousands of hectares, many of them in the hands of smallholders, are now planted with rice strains made more resistant to lack (or excess) of water or, alternatively, to salinity, under the supervision of the International Rice Research Institute based in Manila. One can indeed speak of the beneficial extensions of natural biodiversity that are made possible by the convergence of natural forces, scientific knowledge, and appropriate institutional mechanisms.

WATER: SPECTACULAR SAVINGS

A GIANT STEP IN IRRIGATION TECHNIQUES

Did a vigorous solitary tree in the middle of a dry patch of land in Palestine inspire the most significant modern innovation in irrigation methods? Did Simcha Blass or Daniel Hillel catch sight of the tree and then have a look underground, discovering a pipe that was leaking drop after drop at the roots of the tree? The answers are somewhat contradictory. It is clear, however, that in the late 1950s and early 1960s both men, working in arid conditions in Israel, pioneered drip irrigation (also called micro-irrigation), taking advantage of another innovation at the time, cheap plastic tubes.

In November 2012, Hillel received the World Food Prize at Iowa State University. In an interview in the *Wall Street Journal* ("Drip, Drip, Drip"), he recalled the early days of drip irrigation in the Negev desert: "We realized through drip irrigation, by applying water to the rooting zone of crops very gradually, drop by drop, the soil is never saturated nor ever allowed to desiccate. Consequently, the system becomes more sustainable, water is used more efficiently, and farmers could get much more crop per drop" (October 15, 2012). Less water and richer harvests—it was indeed a revolution in agriculture.

Blass had his own system patented in the traditional way, whereas Hillel made his system freely available, contributing personally to the adoption of drip irrigation in more than thirty countries, including such neighbors as Jordan and Egypt. Drip irrigation is by no means convenient for all crops; even for suitable crops, it is not universally adopted by farmers. However, it is practiced in countries on all continents, including parts of China, India (the largest performer), and the United States. Drip irrigation systems can be very simple and inexpensive, using basic components that can be made from recycled materials and that function on Newton's gravity; in this way they are affordable to poor farmers in countries like Bangladesh, Burma, and Zimbabwe. Of course, they can also be sophisticated, with all sorts of monitoring devices.

Technical sophistication is not an aim in itself, but it may be efficient in suitable circumstances. Farmers in Nebraska use the water they extract from the Ogallala Aquifer in a radically different way from the farmers in the High Plains of Texas (see chapter 2): the Nebraska farmers put to good use a sophisticated irrigation system designed by the German engineering firm Siemens. Sensors scattered across the fields capture information about soil moisture on the ground and about temperatures at the top of plants; this information is transmitted to a central computer that processes it and then fine-tunes the amount of water delivered to the plants. Pioneered by a few farmers in central Nebraska, the system has been tested and validated by experts in irrigation from the College of Agricultural Sciences and Natural Resources at the University of Nebraska–Lincoln. It has since been picked up by farmers all over the state, with the university acting as mediator in adoption.

FROM WASTEWATER TO POTABLE WATER

We have just seen how designing with combined nature and technology is bringing spectacular water savings in agriculture. Purifying and recycling domestic "wastewater" (in the sense of already-used water) is another area with great savings potential. Progress in such systems is best obtained by designing with both nature and technology, sometimes based on rather advanced science. Even if the result is not potable water, it is usually adequate for agricultural and industrial needs, as can be seen in San Antonio—the second-largest city in Texas—albeit more on golf courses than in agriculture.

Producing potable water from wastewater is becoming standard recycling practice in many places.[4] Recycled water is used by households downstream of central Paris, in Singapore, and in Orange County, California (note that this list is not exhaustive). Looking at why and how the local government and residents of Orange County decided to adopt recycled water makes an interesting case in the design and adoption of innovation.

Orange County sits on the Pacific coast between Los Angeles County and San Diego County. The county's 2.3 million inhabitants (up from 700,000 in 1970) were completely dependent for their water supply on inflexible—and costly—allocations of water imported into Southern California from the Colorado River system and from the Sacramento–San Joaquin River delta. Without any prospect of having allocations increased, without significant rainfalls, and without reserves left in a depleted aquifer, the county authorities seriously looked into both recycling and desalination, with the objective of producing approximately a third of the county's potable water. They chose recycling on the basis of its large cost advantage. In the process, the water would be purified in a technologically advanced plant and then injected into the aquifer. Injection would have the dual advantage of being an additional step in the purification process and of contributing to block seawater intrusion into the aquifer.

Extensive information was provided, and discussions were organized all over the county for up to three years, while the necessary technical studies were completed and the administrative requirements fulfilled. The cost of the recycled water represents roughly two-thirds of the price paid for imported water. A beneficial side effect of recycling is the reduction of the discharge of untreated wastewater into the ocean. The EPA granted Orange County Water District the 2008 Water Efficiency Leader Award (category Government) "for success, and in fact world leadership, in wastewater purification for groundwater replenishment. Recognition is given for technical success but especially for the effort to educate the public from day one."[5]

Not everyone was happy with Orange County's decision. It was greeted with horror in neighboring counties, particularly in San Diego

County, which had the slogan *San Diego should flush "toilet to tap" plan.* However, faced with the consequences of a persistent drought all over California, San Diego has since reassessed its position—a first purification plant is scheduled to be operational there in 2021.

SOIL: BIOLOGY FOR SUSTAINABILITY

UNSUSTAINABILITY IN AGRICULTURE

World-class experts, national academies, and international institutions concur: the way soils are mishandled around the world is unsustainable. In developed and emergent countries, this situation stems from uniform reductionist approaches to agriculture, dominance of chemistry over biology, and astounding levels of resource waste, all of which are supported by governments that spend more than $200 billion annually in public agricultural subsidies (Organisation for Economic Co-operation and Development 2016). In developing countries, poverty often consigns soils to exhaustion. These experts and institutions, however, show that reorientations are possible, promising, and urgent.

For example, in its 2008 *Development Report*, which is entirely devoted to agriculture, the World Bank raised the alarm: "Equally alarming has been mounting evidence that productivity of many of the intensive systems cannot be sustained using current management approaches. There is growing evidence that soil degradation and pest and weed buildup are slowing productivity growth" (188). In the summary introducing its *Report on Science and Agriculture*, the Royal Society (2009) shared the diagnosis and pointed to remedies: "Current approaches to maximizing production within agricultural systems are unsustainable; new methodologies that utilize all elements of the agricultural system are needed, including better soil management and enhancement and exploitation of populations of beneficial microbes. Agronomy, soil science and agroecology—the relevant sciences—have been neglected in recent years" (ix).

To reduce the detrimental effects and the waste of resources associated with conventional industrial agriculture, a number of farmers in developed and emergent countries (India, in particular) have turned to a set of practices known as *precision farming*. They strive to apply tightly calibrated quantities of chemical implements at the right place and the right time; water inputs and tillage are also well adjusted. As a result, encouraging results have been obtained.[6] This approach doesn't aim to mobilize biological interactions and dynamical processes for agricultural purposes, though such mobilization is the key to sustainable resource use and yields.

In a biology-based approach, there are a great number of techniques for a diversity of needs and circumstances. In the limited space here, we consider two techniques that are particularly significant as far as conservation and enhancement of soil qualities are concerned. Keep in mind that they are just two in a large family that grows with growing research, development, and promotion efforts, which, for too long, have been dwarfed by the resources poured into the development of chemical fertilizers and biocides.

BYPASSING CHEMICAL FERTILIZERS AND BIOCIDES

Interspersing legumes and fertilizer trees (which fix N_2 from the air) with crops like wheat or maize, or planting them in rotation, is an efficient and generally nonpolluting way to bring back to the soil the nitrogen taken away by crops. Moreover, the efficiency in fixing N_2 may be enhanced by appropriate breeding techniques. This approach has additional benefits: legumes and trees provide a perennial cover that reduces erosion; they reduce the germination of weeds as sunlight is filtered; and they reduce the circulation and development of pests specific to the main crop grown in the field.[7] With their more extended root systems, perennials can access nutrients and water beyond the reach of the crop with which they coexist and cooperate.[8]

Rotations with legumes have been practiced on a large scale in Australia for the past 50 years, sustaining significant increases in cereal yields. In Southern Africa, "fast-growing fertilizer trees such as *Gliricidia*, *Sesbania*, and *Tephrosia*, have improved soil fertility, soil organic matter, water infiltration, and holding capacity" (World Bank 2008, 164). In a developing country like Malawi (Gilbert 2012)—where the soil has, in many places, been exhausted by maize monoculture without proper replenishment of nutrients—legumes (pigeon peas, groundnuts) and fertilizer trees have made it possible for smallholders to increase extremely low yields (1 ton/ha of maize) to more rewarding ones (3 tons/ha). The farmers also collect significant amounts of legumes with high protein contents. The wood secured for domestic uses is also a valuable resource. Yet these results cannot be obtained without a significant increase in labor effort and in the diversity of tasks required.

In the ongoing fight against pests, soil biodiversity is all too often the collateral victim of chemicals applied indiscriminately. Not so with the push-pull system researched and developed as a joint project at the International Centre of Insect Physiology and Ecology, Kenya, and Rothamsted Research, England (International Centre of Insect Physiology and Ecology 2011). The targets of the system are maize stem borers, which attack maize (the main crop in East Africa) from within the stems. The insects are "pushed away" by trees intercropped with maize, and they are "pulled" (attracted) to surrounding grass, which also attracts some of its predators. (For a detailed analysis of the design and dissemination of this innovation, see chapter 8.)

These are only two examples—albeit significant ones—of biologically integrated agricultural approaches. There are many more on the ground, attuned to a huge diversity of specific circumstances. Jules Pretty et al. (2006) reviewed a rather large number—12.6 million farms in fifty-seven countries—which actually covers only 3 percent of land cultivated worldwide. If we want to feed nine billion people by 2050, the pace of adoption of such approaches should be seriously ramped up.

Availability of agricultural practices—however sound, broad, or adapted to the specific conditions of application—is not enough to guarantee sustainable uses of soil. Such practices cannot be meaningfully implemented without being embedded in legal, economic, and social conditions that protect farmers and support their efforts (see chapter 8 on advances in Kenya and Tunisia). In this respect, World Bank (2008) offered a clear review of the main relevant issues.

ENERGY: EFFICIENT PRODUCTION AND SOBER CONSUMPTION

BREAKTHROUGHS IN ELECTRICITY STORAGE

The efficiency of wind and solar energy production and uses could be boosted by significant improvements in energy storage techniques, as the problems associated with intermittency, in particular, could be significantly alleviated. However, but for terrestrial, two-level pumping stations, most of which are located in mountainous regions, high-capacity storage devices are not yet available. Marine pumping stations might become a useful addition, but for the time being, there are only working prototypes in Japan and advanced projects along the North Sea coast.

There have been breakthroughs in flow batteries, which essentially comprise a cell, in which chemical energy is converted into electricity in a reversible reaction, and two tanks filled with two matching liquid electrolytes. The electrolytes are simultaneously pumped through the cell, where they react to produce electricity. After being regenerated, the electrolytes are again pumped through the cell. Thus, recharging a flow battery boils down to bringing in tanks with regenerated electrolytes. The storage capacity of a flow battery is flexible; a relatively low capacity would meet the needs in a home, a far larger one would be needed for serving a community. The cost of current flow batteries is prohibitive, and some of their ingredients are toxic (Perry 2015). However, a team based at the Harvard School of Engineering and Applied Science has innovated a way to use organic components to make their battery both safer and significantly cheaper than those currently available (Lin et al. 2015). A different model is being developed along the same lines at the Pacific Northwest National Laboratory. Similarly, at Argonne National Laboratory, a program is underway that is specifically geared toward devising and developing innovative and economically competitive electrolytes—in particular, "nanoelectrofuels," or liquids containing electroactive nanoparticles in suspension—which offer high energy density at a reasonable cost. These developments should significantly enhance the potential of wind and solar energy, reducing the need for backup coal- or gas-fired power stations.

Decentralized production and consumption of electricity—mainly, solar—and the convenience of electric vehicles will also benefit from enhanced and cheaper solid-state batteries, such as the lithium-ion batteries soon to be produced by Tesla on a large scale.[9] More breakthroughs in the field of lithium-ion batteries are in the offing, including a sealed battery (minimizing instabilities) with enhanced volumetric energy density, developed at MIT.

Recent R&D results with supercapacitors might offer an alternative to the more usual chemical batteries. Electricity storage in a supercapacitor does not depend on chemical reactions; instead, it is realized within suitable nanomaterials (e.g., carbon nanotubes) as static electricity, which makes for much faster recharges than are possible in chemical batteries. A South Korean team at the Gwangju Institute of Science and Technology claims to have significantly improved the performance of supercapacitors by using ultrathin sheets of graphene to store the electrons: "This is a remarkable leap in energy density for supercapacitors and puts storage capacity parity with Li-ion batteries within reach. And when you have a supercapacitor that has the storage capacity of a Li-ion battery and the ability to charge in mere moments rather than hours, all-electric vehicles might just be a lot more attractive" (Johnson 2015, 2). It is also worth stressing that, whereas lithium-ion batteries depend on the availability not only of lithium but also of some rare metals, supercapacitors are not dependent on lithium and significantly less so on rare metals.[10]

A VARIETY OF PATHS FROM PHOTONS TO ELECTRONS

The most promising source of renewable energy is sunlight, and the most promising tool for realizing the potential of sunlight is the photovoltaic cell.[11] Large numbers of subsidized bulky silicon panels ("wafers") have been (and continue to be) installed worldwide. The future of solar electricity, however, lies in a wealth of innovative devices, from which a small number will emerge along with appropriate managerial and economic approaches to their insertion in the economy.

It is possible to print cells on thin films with conversion rates (from photons to electrons) above 20 percent in operation (i.e., not merely in lab tests), compared to 13–15 percent for well-performing panels. However, these films still need to be protected within transparent frames and installed by skilled technicians; the corresponding costs are greater than the cost of producing the cells themselves. To make photovoltaic electricity competitive without subsidies, all these cost components must be trimmed. The results of such trimming may be significant, illustrated by the fact that the difference in prices in France and Germany is almost entirely due to lower costs of installation in Germany. First Solar, a U.S. subsidiary of the French oil major Total, is not only the industrial leader in thin films, it is also involved in installing and financing systems. SolarCity, another U.S. company, doesn't produce any panel or film; it instead specializes in installation and financial services. For example, it offers to keep the ownership of a system it has installed on a roof. The owners of the roof pay for the electricity they consume at competitive rates and have the option to buy the system at conditions specified in the initial contract.

A third generation of solar technology is now in sight—for instance, cells made of organic materials can be "painted" on wrinkled surfaces, roofs, and walls (Yu, Zheng, and Huang 2014). For especially demanding uses, when high yields are worth high costs, conversion rates above 40 percent are reached by superimposing layers made of different materials that catch light at different wavelengths. Many other innovations are likely in the pipeline, as both the physics and the economics of light conversion are remarkably versatile.

The high potential of these various approaches justifies serious efforts of research and preindustrial development, with significant public and private support (which should no longer go to outdated technologies) to help diversify uses and bring costs down (Philibert 2011).[12] The sun provides unlimited (at least in terms of human consumption) and broadly distributed energy; it is indeed worth the effort to tap it efficiently.

AN UNEXPECTED SOURCE OF BIOFUEL

As far as advanced biofuels are concerned, they could be produced from algae that feed on CO_2; yields might be high, and very little land, if any, would be required. However, costs of production are still far too high. Research on this process is actively underway, particularly in California and Japan. Also in California, at the J. Craig Venter Institute, researchers are trying to synthesize (a potential application of "synthetic biology") new elementary living organisms optimized for transforming CO_2 into fuel.

Among the travails of the so-called second-generation biofuels (see chapter 3), an outsider has popped up with possibly nearer-term prospects than algae and synthetic organisms. Its credentials were enhanced by the display in *Nature Communications* (Wang et al. 2014) of the complete genome of *Spirodela polyrhiza*, or greater duckweed. *Science Highlights* (February 21, 2014) hailed the sequencing of *S. polyrhiza* as a breakthrough: it "provides clues about how the tiny plant can be used as an efficient biofuel raw material. . . . Understanding which genes produce which traits will allow researchers to create new varieties of duckweed with enhanced biofuel traits."

Duckweed is the smallest flowering plant and is present at all latitudes; the plant looks like a tiny green lentil floating on still waters such as ponds and small lakes. Its asexual mode of reproduction—clones are produced from leaves—makes it the fastest-reproducing flowering plant. Because duckweed feeds on nitrogen compounds and phosphates from municipal sewage or agricultural runoffs, it is a vector of bioremediation. Collecting duckweed plants to exploit their other potential use—that is, the production of biofuels—would make them ubiquitous vectors of bioremediation. What makes them good at producing biofuels is the combination of the following:

- A high starch content and, to a lesser degree, a high lipid content, which are suitable inputs in the fermentation process leading to the production of biofuels
- A low lignin content (with lignin being an impediment in this process)

The extended genomic information on duckweed is the basis for devising genetic manipulations that aim at even quicker reproduction and that ensure a higher starch and lipid content and less lignin. The process involves more than standard genetic manipulation; indeed, epigenetics has been mobilized: genes that command the production of starch or lipids will be activated, while genes that command the production of lignin will be silenced.

ENERGY EFFICIENCY AND ENERGY SAVINGS IN CONSUMPTION

Efficient patterns of energy consumption are no less important than efficient modes of production. Aiming at ever-increasing production while disregarding how the energy produced is used and how much is wasted might be sustained for a while, but not in the long term; the negative side effects of both energy production and consumption pile up and become unbearable. Until now, energy waste has been tolerated to such an extent that it is at once necessary and possible to drastically cut it down. Waste appears as both a nuisance and a resource. To make use of that resource, it is essential to switch to a path of technical, organizational, and behavioral innovations in consumption. We will illustrate the dynamics involved in two sectors: heat and electricity efficiency in buildings and mobility systems for people and goods.

Heat and Electricity Efficiency in Buildings

It is technically possible to design and realize new buildings, whether residential or professional, that require no net energy input for heating and air conditioning. In Northern European countries, where numerous completed projects have bred knowledge, it is indeed crucial to have architects, engineers, and craftspeople well versed in the new techniques and convinced of their value and, of course, economies of scale. In such conditions, zero-energy import designs come at no extra building costs.

Efficiently retrofitting existing buildings, such as apartment buildings in London, New York, or Paris, is also possible; efficient thin insulation panels are now currently available, as is window glass with embedded nanoparticles that block infrared radiation, keeping heat inside in winter and outside in summer. Having such conservation devices justifies installing heat-extraction tools (e.g., heat pumps) and automation systems monitoring and controlling lighting, ventilation, heating, and air conditioning.[13] Retrofitting existing buildings, however, entails investment costs that are not recovered quickly; rather, they are paid for by the energy savings that they make possible. The time it takes depends on the quality of the improvements realized and on local circumstances. Such upfront investments often deter individual owners and businesses. Hence, such practices might instead require public subsidies—which are often unavoidable as part of a policy of social housing renovation—and might also call for and induce organizational innovations, with firms that provide materials or builders asking for only a partial upfront payment and then sharing in the benefits of energy savings (see earlier for similar arrangements in the solar electricity sector). Such innovations in financial arrangements are implemented in Germany and more generally in Northern Europe, where waste heat is recycled on a large scale in district heating systems.

As far as electricity is concerned, its provision will, for some time yet, mainly come from centralized means of production; thus, consumption and production will have to be continuously adjusted as long as large storage facilities are not broadly available. In such circumstances, modulating electricity demand has the potential to significantly reduce production capacity requirements. Many countries have modulation schemes for industrial demand, but modulating household demand on a large scale is another matter. Advanced information and communication technology, on one hand, and concern for users' explicit acceptance, on the other, have made successful demonstrations possible in the United States, Canada, France, and Germany, to name a few. The main (though not exclusive) objective is to erase peak demands that otherwise require extra capacity that is running for fewer than fifty hours a year. In fact, reducing demand during peak hours appears to be one of the most efficient energy-saving measures.

In this context, some utilities have instituted "demand response programs," such as those offered by CPS Energy in San Antonio

(Texas), which is the largest municipally owned U.S. energy utility. Other firms (Enernoc, Entelios, Kiwi Power) specialize in the production of demand cuts. A typical contract offered to interested consumers specifies the cuts that the customer allows the operator to perform; electrical appliances may then be operated from the firm's headquarters, relying on a network of computers continuously confronting total demand and production in order to determine the relevant amount and distribution of cuts. To boost consumers' willingness to sign on to such a project, the firm provides a command that users can use to suspend a cut. The operator is paid from the savings in production and transport capacities.

Saving electricity should be pursued by all effective means, especially those mentioned here, as it is the cheapest way of reducing energy-based CO_2 emissions.

Mobility Systems for People and Goods

There are two main ways to change how mobility is practiced without unduly cutting down on it:

- Technical: improving fuel efficiency, switching to non-fossil fuels, designing and using dedicated vehicles, promoting economies of scale in public transport equipment, and so on
- Organizational and behavioral: mixing and articulating different patterns of ownership, making different transport modes communicate between them and with consumers to facilitate their joint use, and so on

Fuel-efficiency improvements have been implemented in many countries, and objectives for further improvements have been set. Differences between countries—the United States is less advanced and less ambitious than countries in the EU, for example—show that the issue is only partly a matter of technology: political choices (fuel taxation, in particular) and consumer preferences (size of vehicles, in particular) also weigh on the decisions.

Hybrid and electric vehicles have the potential to displace fossil fuels to a significant extent, at least in places where electricity is mainly produced from non-fossil fuels; thus, the potential may be realized to a larger extent in France, where 90 percent of the electricity is nuclear or hydro, than in Germany, where coal is still dominant. To minimize the vehicle's energy requirements and to maximize convenience for users, Michelin has a prototype of a four-passenger electric car with the engine and brakes distributed between the four wheels (an idea that Ferdinand Porsche first experimented with in 1900) and with flat batteries under the floor. The gains in stability of the vehicle and in room for people and luggage are striking. Better stability and an orientation toward urban use make it possible to reduce the weight of the car, thus reducing energy consumption and energy storage. The Michelin car is dedicated and very well adapted to urban use.

Dedicated vehicles are no novelty as far as the transportation of goods is concerned: on the "last mile" within cities, rather than heavy long-haul trucks, one sees vehicles adapted in size, equipment, and motorization to the various demands they serve. But who would buy a dedicated car? Formulated in this way, the question is far too narrow and ignores the many ways car ownership may be shared, modulated, and delegated. Consider a resident in Paris. She would not own a car; instead, she would rent an electric Autolib' (a joint project between investor Vincent Bolloré and the city of Paris), available for instant hire throughout the city, whenever she has a trip to make within the city or the adjacent suburbs that would not be convenient by metro or bus. For longer trips, in France or elsewhere in Europe, she would combine fast train connections with rental car services on the basis of a contract for repeated occasional hires; car-sharing would also be an option. A resident in faraway suburbs would have fewer opportunities, though public transport might be available at a walking distance from his home. If not, he might own a city car that he would use on workdays to reach the next train station and on weekends for various local uses. If he does not prefer (or have access to) public transport, he could share a larger car with neighbors working in the same inner-city district.

To make the most of such combinations of transport modes and ownership patterns, consumers should be informed in real time and with clarity of all available offers. In an era of more mobile phones than toilets, is it completely unrealistic to expect that information/communication-savvy countries like China, India, or Brazil could promote not only the technical but also the organizational and behavioral innovations involved in such a change?[14]

A MANHATTAN PROJECT FOR CONTAINING CLIMATE CHANGE

WHY NEGATIVE CO_2 EMISSIONS?

Within the next twenty years, it could prove a fatal peril to go on increasing the stock of GHGs in the atmosphere (see chapters 3 and 4). Yet, humankind seems unable and unwilling to control as tightly as required the ever-increasing volumes of GHG emissions. How to avoid being squeezed? The only way that doesn't itself tinker with essential climate mechanisms—as geoengineering projects would (Barrett 2012)—is to pump CO_2 at concentrated sources or from ambient air and then dispose of it in the most appropriate ways. Pumping and storing might not be cheap; thus, driving costs down is worth R&D efforts on par with the Manhattan Project, as the stakes are even higher.

It is often argued that switching decisively to a sustainable development trajectory requires a "Manhattan Project" or an "Apollo Program" of sorts. As far as globally sustainable development is considered, Richard Nelson, who along with Kenneth Arrow started the economics of innovation, strongly disagrees: he is adamant that development stems from a broad diversity of decentralized endeavors (see chapter 8). Nelson, however, argues that there is one domain of R&D that calls for a Manhattan Project, and that is CO_2 capture (Sarewitz and Nelson 2008).

ROCKS THAT NATURALLY CAPTURE CO_2

Nature itself captures CO_2 from ambient air. Silicates of calcium or magnesium are significantly more reactive with ambient CO_2 than are other minerals. The reaction forms stable solid carbonates—hence, the name *mineral carbonation*. These elements are found at higher densities in two kinds of rocks—basalt and peridotite. Natural weathering of such rocks fosters mineral carbonation. Peridotite is particularly productive; unfortunately, while abundant underground, it is rare on the earth's surface, with a significant presence restricted to the Sultanate of Oman, Papua New Guinea, New Caledonia, and the east coast of the Adriatic Sea. (Oman contains about 30 percent of the area where peridotite is apparent.) According to geologists Peter Kelemen and Jürg Matter (2008), from the Earth Institute (Columbia University), who have extensively investigated Oman: "In situ carbonation of peridotite could consume more than 1 billion tons of CO_2 per year in Oman alone, affording a low-cost, safe, and permanent method to capture and store atmospheric CO_2" (17295). This presupposes fracturing the rock to increase the contact area between air and rock.

Humans also might help: "Engineers have much to learn through understanding how natural processes achieve almost 100 percent carbonation of peridotite" (Kelemen et al. 2011, 546). In this case, understanding is not an objective in itself; the ultimate objective is to find ways of boosting carbonation of peridotite and basalt, either with CO_2 in ambient air or CO_2 dissolved in seawater, as basalt is relatively abundant on and beneath the sea floor.

Promising preliminary results have been obtained in Iceland. The CarbFix Project, started in 2012 near Reykjavik, has indeed shown the capacity of underground basalt to turn CO_2 into stable solid calcite and magnesite. CO_2 dissolved in water is pumped into underground basalt, and the reaction produces carbonate veins in the rock. The good news has been the speed of the process: 95 percent of the CO_2 injected solidified within two years, whereas at least a decade was expected. The downside is the water requirement—25 tons of water per ton of CO_2 injected—though further technical progress might reduce that ratio. Alternatively, the operation might be replicated in shallow seawater, as salt in the injected water is no impediment. Costs depend on local conditions, from $30 in Iceland up to $100–$150 anticipated in less favorable conditions. The project is a collaborative endeavor between The Earth Institute at Columbia University, University of Southampton, University of Copenhagen, University of

Iceland, and Reykjavik Energy. An assessment of the results was published in *Science* (Matter et al. 2016).

ARTIFICIAL DEVICES FOR PUMPING CO_2 FROM AIR

There are other technologies for capturing CO_2 either from gas flows in power plants or from ambient air. After a string of cancellations or inconclusive attempts in the United States and Europe, Boundary Dam coal-fired power plant in Saskatchewan (Canada) has been retrofitted at a cost of $1.2 billion to separate the outgoing CO_2 flow from the flue gases. The CO_2 is then piped into a nearby oil field. Operations started on October 1, 2014, and immediately aroused interest worldwide. The capture technology used at Boundary Dam is, however, heavy on energy. Exxon and the specialized firm FuelCell Energy are working on a new technology aiming at recovering a significant fraction of that energy by feeding the flue gases into fuel cells.

As far as capture from ambient air is concerned, U.S., Swiss, and Swedish teams are working on artificial materials that might prove efficient at capturing CO_2.[15] One method is to absorb the CO_2 on a specially designed plastic surface; when the collector is saturated, CO_2 is discharged, and the collector goes back to work absorbing more CO_2. Klaus Lackner, who has pioneered the technology, stresses, "Air capture of CO_2 may provide an option for dealing with emissions from mobile dispersed sources such as automobiles and airplanes" (Lackner et al. 2012, 13156).

When streamlined, these CO_2 capture technologies would entail a $30–$100 cost per ton of CO_2, according to their proponents, which one might compare with the taxes imposed in Nordic countries and in British Columbia (up to $150 in Sweden; see chapter 9). Although these cost estimations are disputed, what is not disputed is that more R&D and demonstrations might greatly improve the perspectives for cutting costs, as is often the case with emerging technologies. It is puzzling that so few financial resources have been made available to develop and deploy capture technologies: less than one-thousandth of the amounts that fossil fuel firms invest annually. As economist Scott Barrett asks, will we wait for "an emergency situation, perhaps after a climate tipping point had been crossed[?] The imperative to reduce the level of atmospheric concentrations, quickly, would certainly commend the use of air capture" (Barrett 2012). Thus, it would be more rational to start a Climate Manhattan Project without further delay.

What to do with the volumes of gas collected in these various ways? Gas can be injected into suitable geologic formations, as is done at the Boundary Dam power plant. However, Boundary Dam is a particularly favorable configuration. Rarely do power plants sit in the vicinity of suitable geologic formations; even when they do, neighboring populations might object, whether for good or unfounded reasons. Capture from ambient air, as its location is not predetermined, is in a more favorable position; to a certain extent, it compensates for the inconvenience of having to pump CO_2 from a more diluted source. With peridotite and basalt, storage comes with capture: the chemical transformation of the rock entails both capture and storage. Being much more porous (and abundant) than peridotite, basalt somewhat compensates with greater storage capacity for its inferior rates of capture.

Relatively small quantities of pumped CO_2 do find buyers right now in the chemical industry. In the future, large quantities might be used to produce synthetic fuels. George Olah, Surya Prakash, and Alain Goeppert (2011) "have developed the concept of the Methanol Economy and some of the underlying new chemistry for using methanol as an energy storage medium, transportation fuel, and raw material for producing synthetic hydrocarbons and their products. It is based on the chemical anthropogenic carbon cycle, combining carbon capture and storage with chemical recycling" (12895). This way of storing CO_2 would not reduce the amount present in the atmosphere, but it would at least close the open carbon cycle associated with the burning of gasoline made from fossil oil.

Although Anders Eldrup, former CEO of Danish power utility DONG Energy, had not heard of Olah, Prakash, and Goeppert's work, he came to concur with them. During the winter of 2009–2010, the Danish wind farms produced such an excess of electricity, which had to be evacuated through the interconnectors to Norway and Sweden, that the price became negative. With the multiplication of offshore wind farms in Danish waters, it is anticipated that imbalances between electricity production and consumption might result in negative prices for about 1,000 hours per year by 2020. Eldrup came to realize that, rather than incurring losses on excess electricity, that excess should be used to produce hydrogen. Similar conditions will prevail in the United States, in places in Texas and the Midwestern states, as wind farms are multiplied. With hydrogen and CO_2 at hand, it is not difficult to produce methanol, a substitute for gasoline, which would also prove a means of recovering electricity in excess and of dealing with the intermittency and volatility of renewables.

The energy requirements of both the U.S. way of life and the economic development in Asia and elsewhere will most probably not be curbed in proportions sufficient to avoid a climate catastrophe. Hence, while strictly containing GHG emissions is vital, recapturing some of them, either at the points of emission or from ambient air, is no less vital. However, neither containment nor capture will be realized on the required scales without strong public policies supporting private and public initiatives—in particular, proper pricing of the greenhouse gases at levels that create sufficient incentives for effective action (see chapter 9).

7

SCIENTIFIC UNCERTAINTY, FABRICATED UNCERTAINTY, AND THE VULNERABILITY OF REGULATION

NEWTONIAN PHYSICS is a world of strictly deterministic causality. Although uncertainty is ubiquitous in quantum physics, it is governed by probabilistic laws that produce statistical determinism. The situation is different in sciences like ecology, climatology, oceanography, and medicine, which are no less critically linked to sustainability. In these sciences, uncertainty is more loosely structured and hence more difficult to deal with; to various extents these sciences are genuinely uncertain.[1] As geophysicist Henry Pollack argued in his 1997 book, *Uncertain Science . . . Uncertain World*:

> Because of the complexity, it is extremely difficult for even the most capable ecologists to study a forest ecosystem in its full detail, and so they develop simplified concepts about the workings of the ecosystem, focusing on a few elements and their interactions that are thought to be particularly significant. This conceptualization of the ecosystem web of interactions is called a model. To be sure, different ecologists may perceive the interactions differently, weigh the participation of the different components differently and, therefore develop different models. Because of the complexity, the ecosystem is imperfectly understood and uncertainty about how it hangs together is attendant. (106)

This uncertainty seems to present an actual chasm. However, for the past twenty-five years, scientists (including economists) have developed structured and rigorous approaches to evaluation and decision making while working under conditions of genuine uncertainty.

We should first stress that the uncertainty discussed in this chapter, however genuine, is partial; a core of certain knowledge is, in general, available. For example, the various models of a forest might be in agreement as far as the fundamental interactions within the ecosystem are concerned. Likewise, the main climate models provide a common body of core results that are scientifically undisputed and that constitute a solid basis for devising relevant climate policies. Nevertheless, there remains a fair amount of genuine, and sometimes irreducible, uncertainty—hence, the importance of developing rational approaches to evaluating and deciding under genuine uncertainty, properly taking into account certain components at hand. In this chapter, we consider two approaches: (1) tests assessing the reliability of a piece of uncertain science and (2) general decision-making procedures for areas where there is genuine uncertainty. These are essentially structured approaches for making the most of uncertain, yet useful, information.

From the asymmetric information models of economists George Akerlof, Michael Spence, and Joseph Stiglitz, we are well aware that a frequent winner is not necessarily the one who is good; rather, it is the one who merely looks good while avoiding the costs of actually being good. Such a strategy has too often been successfully implemented in disguising science—and indeed fabricating uncertainty—be it the science of the health hazards linked to tobacco or various pollutants, the science of the ozone hole, or the science of climate change. These manipulations have played on the unease of both the public and politicians regarding a science that, being genuinely uncertain to a certain extent, is mistakenly perceived as unreliable—the more so as some people and organizations are busy inflating that uncertainty. Former prime minister Margaret Thatcher, as a scientist herself, was not fooled: "The need for more research should not be an excuse for much needed action now" (UN Climate Change Conference, Geneva, 1990). But consider the recommendations, formulated by pollster Frank Luntz in his "Straight Talk" memo, for the benefit of

Republican candidates (including President George W. Bush) in the 2004 elections, as leaked to the *Guardian* in 2003:

> Voters believe there is no consensus about global warming within the scientific community. Should the public come to believe that the scientific issues are settled, their views about global warming will change accordingly. Therefore you need to continue to make the lack of scientific certainty a primary issue in the debate. ("Memo Exposes Bush's New Green Strategy," March 4, 2003)

In the wake of the 2016 elections, some among the most ardent deniers actually joined Donald Trump's administration.

When it has been recognized that a domain of activity must be regulated, industries like the chemical industry do not hesitate on waging a war on science (and on regulatory authorities). Uncertainty is then a weapon of choice, contributing to the vulnerability of regulation. Uncertainty in science indeed bears heavily on the conduct of regulation, particularly in matters of health and environment, where regulation is most needed. How far obstruction and manipulation might go, sometimes bordering on criminal activity, is well illustrated on both sides of the Atlantic by the pressing need and the seeming impossibility of regulating "endocrine disruptors" (chemical agents that interfere—dramatically, as concerns reproduction—with the hormonal system in mammals). It is difficult to imagine anything more unsustainable than an activity that viciously attacks babies in their mothers' wombs. Yet, it has been sustained for decades (and still is) as a profitable activity by an industry that manipulates science and public authorities on both sides of the Atlantic.

STRUCTURED FRAMEWORKS TO DEAL WITH GENUINE UNCERTAINTY

RELIABLE UNCERTAIN SCIENCE

Consider for a moment the scientifico-political mess the British government faced some thirty years ago. In 1986, bovine spongiform encephalopathy (BSE) was registered for the first time in Welsh and English cows. Then the disease spread at an astonishing pace. But according to the government, "British beef is safe," because no contagion to humans is possible thanks to the "barrier between species." This pseudo-scientific concept, more or less relevant for some other ailments, had no theoretical foundation and, of course, had not been experimentally tested in the case of BSE. The test did not come until 1991, when scientists at the University of Bristol succeeded in inoculating a cat with BSE; as is well known, just one counterexample destroys a hypothesis (see chapter 5)—in this case, the existence of a barrier between species against BSE. The British government and its experts were left without any science, not even uncertain science, at a moment when they had to face a more sinister threat: a new form of the Creutzfeldt-Jakob disease (CJD) striking young people (the original form appears at older ages).

Both BSE and CJD are degenerative diseases of the brain. Could they be linked? A doctor in medicine and biochemist, Stanley Prusiner, happened to be working on related questions at the University of California at San Francisco. More exactly, he was working on neurodegenerative diseases that seemed to result from disorders of protein conformation: it seemed that proteins of the prion type could, after some mutations, become infectious agents in the brain. Never before had anyone observed proteins turned into infectious agents. Dr. Prusiner had nevertheless gathered solid experimental results on mice and had elucidated some of the molecular events responsible for the mutations of the prions and for their infectious power. The foundations were by no means complete—the experimental data were still limited, and not all the relevant molecular events had been elucidated. However, they appeared to constitute a significant subset of what would be a completely satisfactory set of experimental data and theoretical links, and they were coherent. What Dr. Prusiner proposed at the time was a piece of nonprobabilistic uncertain science. However, it was sufficiently reliable (on the basis of its rigorous conformity to the scientific method and of the results already obtained) to scientifically support the decision to bar English beef from being consumed in Britain and the rest of the European Union.[2] It was indeed a piece of *reliable uncertain science*, and as the asbestos disaster has illustrated (see European Environment Agency 2002), it would have been nonoptimal—to say the least—to wait for scientific certainty before making the appropriate decisions.

As far as climate and climatology are concerned, it appears that the IPCC's investigation process contributes to the organization and dissemination of theoretical and empirical knowledge on climate change in a rigorous, systematic, and, in general, well-monitored way. However, there have been some deviations that are indeed regrettable, though they do not diminish the significance of the results made available. Deviations, even far more serious ones, are not that exceptional in science; however, they rarely have serious consequences, because intense competition provides a diversity of ways to produce and communicate important results.[3] Physicist and historian of science John Ziman characterized scientific credibility as follows: "The credibility of science depends as much on how it operates as a collective social enterprise, as it does on the principles regulating the type of information that this enterprise accepts and transforms into knowledge" (Ziman 2000, 58). It seems that the work of Groups 1 and 2 of the IPCC is favorably placed with respect to these criteria, and their findings constitute a reliable, yet partly uncertain, scientific basis justifying far-reaching decisions with respect to climate change. However, Ziman added a warning, reminiscent of Thomas Kuhn's *The Structure of Scientific Revolutions* (1962): "Normal science is a mindset that can take hold of researchers in any field of academic science" (201). Is it possible that the normality of science embodied in the rules according to which the IPCC integrates scientific results has made the IPCC reports slow to mention some relevant manifestations of climate change? This indeed adds to uncertainty, albeit certainly not in a direction that would weaken the IPCC's pronouncements on the necessity of significant and urgent decisions regarding climate change.

GENERALIZING AN APPROACH TO RISK INTO AN APPROACH TO UNCERTAINTY

It thus appears possible to structure decision making under uncertainty according to the tenets of the scientific method, albeit not to the degree of accuracy made possible with the instruments for dealing with risk, as provided by John von Neumann, Oskar Morgenstern, and Leonard Savage (von Neumann and Morgenstern 1944; Savage 1954). The range of applications that stem from their contributions is stunning and can be found in finance, insurance, appraisal of investments, resources, and environment economics. Within economics—and, more largely, within social sciences—it has been a scientific revolution.

Subsequent developments induced some mathematicians and economists to find ways to enlarge the scope of the von Neumann–Morgenstern (vNM) approach, while still having a manageable model encompassing a large range of behaviors. Among the pioneers in this direction are Kenneth Arrow and Leonid Hurwicz. In Arrow and Hurwicz (1977), the authors considered a surprisingly simple, albeit challenging, context: "There are two coins, but coin two is flipped only if coin one shows tails." This is indeed a context that one encounters in subsequent papers about decisions under genuine uncertainty (Klibanoff, Marinacci, and Mukerji 2005; C. Henry 2010; Etner, Jeleva, and Tallon 2012 also provided a useful survey).

Instead of relying on one exogenously given probability distribution, as in the vNM model, the decision maker deals with the information on the basis of an endogenous set of probability distributions. Some of these distributions put more weight on pessimistic assessments of the situation at hand; others put more weight on optimistic assessments. In the *Stern Review* (see section 2.5 in Stern 2006), this fundamental result is interpreted in the following way:

> The decision-maker would act as if she chooses the action that maximizes a weighted average of the worst expected utility and the best expected utility, where best and worst are calculated by comparing expected utilities using the different probability distributions. The weight placed on the worst outcome would be influenced by concern of the individual about the magnitude of the associated threats, or pessimism, or possibly any hunch about which probability might be more or less plausible. It is an explicit embodiment of "aversion to uncertainty," and an expression of the precautionary principle.

It is interesting that the global reinsurance companies, such as Swiss Re, Munich Re, and Partner Re, use methods of that kind to write contracts on events for which no meaningful statistical series is available. So do specialized rating agencies in the United States, the main activity of which is to assess the portfolios of contracts sold by insurance and reinsurance companies. In both cases, it has been recognized that previous practices relying on just one (so-called) mean distribution are misleading, because in so doing, a huge amount of information (uncertain, though valuable) is lost. It is a general proposition that dismissing uncertain knowledge amounts to losing valuable information, which is all the more inefficient because it is possible to deal with uncertainty in a structured way.

As far as climate change is concerned, recent results show that the aforementioned formalized approach to deciding under uncertainty might significantly change the benefit-cost appraisal of mitigation. In "Ambiguity and Climate Policy," Antony Millner, Simon Dietz, and Geoffrey Heal (2013) introduced this nonprobabilistic, though rigorous, approach into the economic evaluation of climate policy. In a dynamic setting, they find that "an increase in ambiguity aversion favors abatement—this ambiguity effect is small when damages are flat, and very significant when damages are steep at high temperatures."

DENYING SCIENCE AND FABRICATING UNCERTAINTY

"Credible science can be translated directly into political power," as science historian Paul Edwards (2010) stressed. This statement is true more generally for science that is perceived as credible. In the seminal model of Akerlof (1970), which reveals the consequences of information asymmetry on the functioning of markets, the largest profit is made by sellers of "lemons" (i.e., bad used cars) when lemons and "peaches" (i.e., good used cars) cannot be distinguished on the market, provided the latter has not collapsed. Similarly, as political scientists Naomi Oreskes and Erik Conway put it in the introduction to their 2010 book, *Merchants of Doubt* (the classic reference on the subject): "Call it the Tobacco Strategy. Its target was science, and so it relied heavily on scientists—with guidance from industry lawyers and public relations experts—willing to hold the rifle and pull the trigger" (6). Such strategy works well as long as its proponents are seen as competent, reliable scientists, and not as producers of unscientific, fabricated uncertainty or as puppets in the hands of manipulators.

THE TOBACCO STRATEGY

In the 1950s and 1960s, more and more epidemiological studies were produced that documented the health hazards associated with smoking. Producers of tobacco—American Tobacco, Benson & Hedges, Philip Morris, R.J. Reynolds, United States Tobacco Company, and others—organized a powerful counteroffensive through a common organism called the Tobacco Industry Research Committee. This committee promoted and sponsored research focusing on the imperfections of the epidemiological studies—exaggerating them when deemed expedient—and on possible alternative origins for the health hazards associated with smoking.

A seemingly scientific institution was created to plan and coordinate this defensive research. Its links with the tobacco industry remained concealed from the public, and its name projected a highly respected image: The Alexis de Tocqueville Institution (François de la Rochefoucauld would see here another *hommage du vice à la vertu*; *Maximes*, 1664, no. 218). In 1976, Professor Fred Seitz, a scientist with an impressive résumé (distinguished solid-state physicist, president for seven years of the National Academy of Sciences and then for ten years of the Rockefeller University, from which he retired in 1979), was put in charge of the research program financed by the tobacco industry. The institution promoted and sponsored both "good cars" (for example, Stanley Prusiner's research on the pathogenic action of mutated prions in the brain) and "bad cars" (spurious statistical studies to put in doubt the validity of epidemiological studies); such a mix of good and bad reinforced the operation's credibility. Many disguised "bad cars" were sold, and the effects lasted a long time. The first national condemnation of the tobacco industry came as late as

2006; based on the Racketeer Influenced and Corrupt Organization Act, the condemnation was indeed infamous, but it didn't erase half a century of exceptional profits on the back of millions of avoidable individual tragedies.

STARS OF UNCERTAINTY

In 1983, when President Ronald Reagan summoned U.S. scientists to support and take part in the Strategic Defense Initiative (SDI, or "Stars War"), a minority led by Edward Teller (a brilliant physicist, who had been involved in the Manhattan Project and later played a pivotal role in the U.S. hydrogen bomb program; he was an inspiration for Stanley Kubrick's *Dr. Strangelove*) enlisted enthusiastically. Fred Seitz was there also, as was one of his tobacco companions, Fred Singer (a specialist in rockets and satellites and a former director of the National Weather Service Satellite Center). Other distinguished volunteers included Robert Jastrow (an astrophysicist formerly associated with the Apollo Program and founder of the NASA Goddard Institute for Space Studies, the director of which, ironically, was later James Hansen, the most outspoken among the U.S. climatologists) and William Nierenberg (professor of nuclear physics at the University of California–Berkeley and former deputy secretary general for scientific affairs at NATO).

Who could imagine that these people were not always behaving as competent and honest scientists? They created the equivalent of the Alexis de Tocqueville Institution, with an equally respectable name—George C. Marshall Institute—which produced a mix of more or less relevant studies, as well as defamatory libels against U.S. scientists opposed to the SDI, with a frequent target being Carl Sagan. These men were animated by a great patriotism, a fierce anticommunism, and a conviction that the state should be strong when it comes to external threats, but inside, the country should not interfere with private liberties and initiatives. As Singer put it: "If we do not carefully delineate the government's role in regulating danger, there is essentially no limit how much government can ultimately control our lives" (Alexis de Tocqueville Institution 1994). This logic helps explain why, when Gorbachev made the crusade to make the SDI obsolete, the institute turned against the environmentalists whom, with his gift for catchy formulas, Singer likened to watermelons—"green outside, red inside."

THE ULTIMATE FRONT: ENVIRONMENT AND CLIMATE

The main instruments and methods had been designed for the defense of the tobacco industry, and SDI had been a rallying cry. Environmental regulations and climate change policy would be the major battlegrounds. The structure of the logistics is invariant, articulating the following actors:

- Behind the scenes are powerful firms (directly or through industry associations) motivated by strong industrial and financial interests. They are much more diversified than in the tobacco case and include ExxonMobil, BP, Shell, Peabody Energy, The National Coal Council, Ford Motor Company, General Motors, and Koch Industries.
- On the scene are fake scientific institutions, established from scratch and financed by the above firms. They have multiplied, always with names that make them look respectable: American Council on Science and Health, Friends of Science, Greening Earth Society, Natural Resources Stewardship Project, and The Advancement of Sound Science Coalition, to name a few.
- Scientists and others are organized and supported within these institutions. First is the old guard from tobacco and SDI, who are still eager to fight and bask in the limelight of the TV talk shows, despite having their scientifically productive life well behind them; they refuse to bow out. There are also newcomers, like meteorologist Richard Lindzen and astrophysicist Sallie Baliunas (highly regarded for her scientific achievements but then spreading the "good word" that the real culprit for climate change is the sun). These scientists are joined by people who are essentially lobbyists, media operators, and manipulators. The most prominent are Mark Morano (who also coordinated "Swift Boat Veterans for Truth," which successfully put in doubt John Kerry's record in Vietnam), Steven Malloy (known as

"junk science" commentator on Murdoch's Fox News Channel), and Frank Luntz (Republican pollster whose "Straight Talk" memo played a remarkable role in the 2004 electoral campaign). Scientists working inside companies like Exxon and Peabody, who were aware of serious risks associated with climate change, have been silenced; these companies are now charged with misrepresenting the issues to stockholders and regulators.

The methods they use have been tested and refined, and they generally work smoothly by making the most of the following:

- *Common sense*: "How can you tell me that they have any idea what climate is going to be like 100 years from now if they can't tell me what the weather is going to be like in four months, or even next week?" (Timothy Ball, from Friends of Science, interviewed on August 12, 2006, by a journalist of the Toronto *Globe and Mail*).
- *Doggedness*: "They promoted claims that had already been refuted in the scientific literature, and the media became complicit as they reported these claims as if they were parts of an ongoing scientific debate" (critique formulated by Oreskes and Conway 2010, 241).
- *Unequal rules*: "Because scientists are so quick to acknowledge when something is not exactly correct, the attackers have won many apologies, corrections or reinterpretations, which they have used to argue that all of climate science is frail and uncertain" (Hoggan 2009).
- *Balance*: "Adherence to the norm of balanced reporting leads to informationally biased coverage of global warming. This bias, hidden behind the veil of journalistic balance, creates both discursive and real political space for the U.S. government to shirk responsibility and delay action regarding global warming" (Boykoff and Boykoff 2004, 134). This is particularly pernicious as a scientific result can now be considered a mere opinion, to be debated against a different opinion.

These investments paid off. According to the Pew Research Center, in 2006, 79 percent of Americans thought that "there is solid evidence the earth is warming" and 50 percent thought it was "because of human activity." In 2010, the corresponding figures were, respectively, 53 percent and 34 percent. In Book 2 of Virgil's *Aeneid*, Aeneas tells Queen Dido the story of Troy's last days:

> Blind with frenzy, we site the accursed creature on top of our sacred citadel. Then Cassandra, who, by the god's decree, is never to be believed by Trojans, reveals our future fate by her lips. We unfortunate ones for whom that day is our last, clothe the gods' temples, throughout the city, with festive branches.

In the United States, Apollo's decree has morphed into ExxonMobil's (and the like's) decree.

What lesson from all this did they draw in China, India, and other emergent or developing countries? Many influential people there came to hold the view that the concern for the climate is essentially a Western conspiracy meant to abort economic development in emergent and developing countries.

NECESSITY AND VULNERABILITY OF REGULATION

First indications appeared in 2001–2002 that compact cars produced by GM—in particular, Chevrolet Cobalts and Saturn Ions—were suffering from serious defects, causing severe accidents. Engines could stall when operating at high speeds due to ignition-switch defects; moreover, airbags were not deploying. The problem bounced around lower levels of the company and dealers, but for years, higher-level managers did not seriously consider the problem. Concerns for safety were not that present on their minds, partly because penalties to deter automakers from ignoring and hiding safety problems were woefully insufficient to motivate them; there was no sufficient incentive to value safety over cost savings. For ten years, in the absence of safety conscience, there was no public communication and no recall, despite hundreds of deaths and thousands of injuries.

During hearings at the U.S. Senate in 2014, the blame was extended to the safety regulator, the National Highway Traffic Safety Administration (NHTSA). It indeed appeared that NHTSA had failed to recognize a systematic pattern in the numerous individual complaints it had received over the years; it had never made the effort to investigate the somewhat complex chain of causes and effects involved in the accidents. Their failure to do so was partly due to a lack of resources and partly to a lack of staff able to keep pace with technological innovations in the automobile industry. Senators had harsh words for both GM and NHTSA: lack of information sharing within and between the two organizations, inversion of priorities, incompetence, negligence, irresponsibility, and so on. Clearly, this situation amounts to serious management and regulation failures, neatly circumscribed and analyzed (albeit, too late), making the causal links transparent and clearly assigning responsibilities. The case had no endless controversies on facts and interpretations, no systematic manipulations of facts, interpretations, and people. There was no significant uncertainty and, hence, no abuse of it. The regulator was simply passive and incompetent.

When uncertainty permeates activities and products, however, regulation gets genuinely more difficult and vulnerable to manipulations—as is the case in the processed food industry. Diagnosing car defects is not marred by uncertainty as is proving that some processed foods have built-in properties that make people feel hungrier after consuming an initial portion (Moss 2014). This is one of the hidden tricks that has contributed to disseminating and amplifying chronic ailments that have become major health concerns in the United States. (Imagine GM engineering defects in the cars it sells.)

When agents from an industry penetrate an institution that is supposed to protect consumers and the population in general, to the point of occupying key executive or scientific positions within this institution, they are able to manipulate facts and rules, and thus morph uncertainty into certainties that promote the industry's interests. This has been the case for the sugar industry (among others), as a broad scientific investigation has shown (Kearns, Glantz, and Schmidt 2015; Kearns, Schmidt, and Glantz 2016), making the United States a world champion of obesity and diabetes.

It is a general proposition that regulation is more vulnerable when dealing with activities and products that interfere with the functioning of living organisms. The dangers involved and the complexity and uncertainty of the cause-effect links are magnified. The case of endocrine disruptors is emblematic. Hormones are biochemical messengers produced by endocrine glands in mammals. These messengers bind to specialized receptors that, translating the messages received, regulate several essential life functions—in particular, reproduction and development processes. When a woman is pregnant, hormones transmitted to the fetus are critical factors for its subsequent development.

Many synthetic chemicals (perhaps about a thousand, though it is impossible to know exactly how many, as new chemicals are not adequately tested before being brought to the market) interfere with the hormonal system, disrupting production and transmission of hormones, as well as execution of messages received—hence, the name *endocrine disruptors*. These disruptors are present in pesticides (Mnif et al. 2011), detergents, resins, processed paper, and plastics (in particular, food containers). Their action is by direct external contact or by ingestion of contaminated food and water; they accumulate in the body—particularly, in fat. They have even been detected in the fat of polar penguins. In human beings, these disruptors are linked to some cancers (Tilghman et al. 2012), diabetes, reduced male fertility, miscarriages, and, most important, damages to the fetus in the mother's womb.[4] Damages appear immediately at birth or later on—in particular, during the development of the central nervous system (resulting in impaired cognitive aptitudes, impaired vision and hearing, and so on). In an official scientific statement, the *Endocrine Society* presented "the evidence that endocrine disruptors [EDCs] have effects on male and female reproduction, breast development, breast and prostate cancer, obesity, and cardiovascular disorders. Results from animal models, human clinical observations, and epidemiological studies converge to implicate EDCs a significant concern to public health" (Diamanti-Kandarakis et al. 2009, 293).

In 1998, the European Parliament adopted a resolution calling upon the European Commission to tackle the issue of endocrine disruptors. It took the commission more than ten years to have a scientific report published (Kortenkamp et al. 2011) that systematically assessed the dangers associated with the endocrine disruptors in matters of health and environment and that defined a framework for formulating regulatory criteria. The report, which was released on December 23, 2011, was unambiguous in its conclusions: evidence of the toxicological properties of the disruptors justifies that they should be dealt with as bioaccumulative and toxic chemicals. The report is also transparent as regards uncertainty within the available scientific knowledge:

> For a wide range of endocrine disrupting effects, agreed and validated test methods do not exist. In many cases, even scientific research models that could be developed into tests are missing. This introduces considerable uncertainties, with the likelihood of overlooking harmful effects in humans and wildlife. (7)

The concerned industry reacted swiftly. They hired the U.S. consultancy Exponent Inc., which, in another context, the *Los Angeles Times* (February 18, 2010) described as a "hired gun": "When some of the world's best known companies faced disputes over secondhand smoke, toxic waste in the jungle, and asbestos, they all turned to the same source for a staunch defense: Exponent Inc. Now the same engineering and consulting firm has been hired by Toyota Motor Corp. as it seeks to fend off claims that sudden acceleration in its vehicles could be caused by problems in its electronic throttle system" ("Toyota Calls in Exponent, Inc. as Hired Gun").

The guns of the industry were then pointed at the European Commission, which was due to propose by December 2013 regulatory criteria informed by the Kortenkamp report. The Directorate General for the Environment was in charge, and a broad meeting was convened at the commission on June 7, 2013, to review the work in progress. The meeting was disrupted by an email from Bayer CropScience to the Commission's General Secretariat, requesting an economic impact assessment before proceeding to regulation criteria. The economic impact would primarily be an evaluation of the impact on the industry's activity. The meeting fell apart.

On July 4, 2013, a noisy "scientific" salvo was fired: a comment by eighteen industry scientists, under the title "Scientifically Unfounded Precaution Drives European Commission's Recommendations on EDC Regulation, While Defying Common Sense, Well-Established

Science and Risk Assessment Principles," appeared in the journal *ALTEX* (Alternatives to Animal Experimentation) and was reproduced in other publications (Dietrich et al. 2013). This title sounds more polemic than scientific, as do all invocations of "common sense" in science; the idea of "well-established science" echoes President George W. Bush's concept of "sound science," and "risk assessment" is woefully inadequate when uncertainty is pervasive. Moreover, *ALTEX* normally focuses on topics not clearly related to the content of the comment: according to its official web page, "*ALTEX* publishes research, meeting reports and news on the development and promotion of alternatives to animal experiments." The comment concludes with the blunt accusation that standards derived from the Directorate General for the Environment's draft proposals would be "contrary not only to science, but the very principles of an enlightened governance and social contract. It is the utmost responsibility of us scientists to resist and counteract any efforts that undermine the core of science and its continuing promise for the betterment of the human condition and of the planet" (382). This message strongly echoes salvos fired during the tobacco and climate wars.

Subsequent publications and confrontations in meetings—including one in the office of professor Ann Glover, chief scientific adviser to the president of the European Commission—proved the comment wrong. With few exceptions, the authors had undisclosed links with firms in the industry or with industry organizations representing these firms. This was particularly true for the lead author of the comment, Daniel Dietrich, who candidly answered interviewers from *Environmental Health News* (September 23, 2013): "We do not believe the discussions on the conflicts of interests will serve anybody because it takes away the focus of the real issue."

Professor Glover nevertheless heeded the call from the industry; she joined the General Secretariat in derailing the process of defining regulation criteria toward the impact assessment initially required by Bayer and then by the whole industry. Members of the European Parliament who had expressed concerns in a letter to the president of the commission on October 16, 2013, were rebuffed in the answer they received five months later (March 25, 2014): "Given the concerns about the possible potential significant impacts on some sectors associated with any particular choice of criteria and the vigorous debate in the scientific community on endocrine disruptors that escalated over last summer, the Commission has decided to carry out an impact assessment."

This response did not include any mention of health and ecological impacts. However, the insistence on a "vigorous debate in the scientific community"—that old recurrent trick—showed that the "Merchants of Doubt" had not yet sufficiently irrigated the European Commission.[5] Will the long-delayed impact assessment at least examine the claim made in the conclusion of Bellanger et al. (2015)?: "EDC exposures in Europe contribute substantially to neurobehavioral deficits and diseases, with a high probability of > €150 billion costs/year" (1256). The corresponding figure in the United States, according to Attina (2016), is $340 billion, or 2.3 percent of the gross domestic product.

In the United States, there hasn't been any coherent attempt at regulating endocrine disruptors. Moreover, as in the tobacco and climate wars, scientists seen as dangerous to corporate or political interests have been methodically harassed. Theodora Colborn, formerly of the University of Florida, was one of the first scientists to investigate and document effects of endocrine disruptors on animal reproduction, specifically working with animals living in or around the Great Lakes. From her experience with the chemical industry, she once advised a younger colleague, Tyrone Hayes of the University of California at Berkeley, to "keep looking over his shoulder, to be careful whom he let in his lab. You have got to protect yourself" (quoted in Aviv 2014). The advice would prove fitting beyond what Professor Colborn had in mind.

In 1999, Syngenta asked Professor Hayes to investigate possible side effects of exposure to atrazine—or, they hoped, the absence of significant ones.[6] Hayes is a specialist in amphibian endocrinology. He discovered that atrazine has devastating effects on the sexual development of amphibians—in particular, frogs. These results were far from what Syngenta had hoped for, and cooperation between the two parties was discontinued in November 2000. However, when Hayes continued working on the subject, Syngenta resorted to all sorts of tricks to "discredit the scientist and destabilize the person" (Aviv 2014).

Years later, in 2012, memos and emails that Syngenta was forced to disclose when facing two class-action suits brought by twenty-three Midwestern cities for "concealing atrazine's true dangerous nature" and contaminating their drinking water indicated that the company had held secret "focus groups" on how to discredit Hayes's results, impede their dissemination, and "mine his vulnerabilities." As one example, other scientists and journalists were paid to disturb his presentations held anywhere in the United States. Syngenta also purchased "Tyrone Hayes" as a search word on the Internet so that searching for him listed supposed reasons why he was not credible. The tricks worked: Syngenta spoiled Hayes's life, and, above all, they influenced the EPA regulator into disregarding Hayes's results.

Hayes's predicament was not an isolated aberration. The tobacco industry, the food industry, the chemical industry, the firms behind climate-change deniers all routinely manipulate facts and interpretations, play on uncertainty, and resort to various harassment tricks to squash critiques and circumvent regulators.

It appears that no regulation in complicated matters is effective unless it is structured on the model of nuclear safety regulation in the countries that have decided to take the issue seriously (see chapter 3). The regulator would then have the power and the obligation to ban any development or product that would reliably (in the sense defined earlier in this chapter) be deemed to carry significant dangers, with appropriate—albeit, not suspensive—appeal procedures.

8

PRODUCING AND DISSEMINATING SUSTAINABILITY-ENHANCING INNOVATIONS

AS SHOWN in previous chapters, without innovation—whether scientific, technical, organizational, or behavioral—there is no prospect of reversing the trend that leads to the devastation of natural capital and the collapse of human societies. However, innovation itself may be destructive; as is the case for science, there is no intrinsic virtue in innovation. It is thus of utmost importance to create incentives and institutions that promote innovations supporting a shift toward a sustainable development trajectory. From creating and sharing knowledge to putting it to use in the vast diversity of local circumstances, this chapter explores ways of effectively supporting and realizing the necessary shift.

Thomas Jefferson (1813), as Isaac Newton before him (see chapter 5), understood that knowledge is a public good. As such, it should be freely available: "He who receives an idea from me, receives instruction himself without lessening mine; as he who lights his taper at mine, receives light without darkening me." Jefferson was nevertheless aware of a dilemma: How do you motivate potential inventors if they anticipate that what they have invented will be made available for free? Joseph Schumpeter (1911) pointed to a way out of the dilemma: give the inventor a patent—that is, a temporary legal monopoly on the products from inventions, though not a monopoly on the background knowledge. The first part of this proposal is the cornerstone of the dominant approach to intellectual property protection. The second part has been forgotten—rights are often granted way beyond the limits Jefferson and Schumpeter considered appropriate. However, this second point is currently being revived in more efficient and sustainable alternative approaches, such as open source.

To foster the production and dissemination of sustainability-enhancing, natural capital–saving innovations, should we mainly rely on making the most of the intellectual property rights (IPRs) as they are currently granted? Or should we consider far-reaching changes in the dominant approach or even the adoption of alternative approaches? Providing meaningful answers first requires us to lay down the traditional legal and economic foundations of IPRs. Then, we will consider how these foundations have more recently been impaired, leaving a situation deeply unsatisfactory and hence in need of substantial changes.

One of the surest paths to reforming the system is to first open the process within which a patent is examined to all concerned parties. Second, important benefits could be derived from introducing more competitive concepts and mechanisms within the fabric of IPRs. Third, it is possible to give more weight to other ways of stimulating innovation, including prizes and guaranteed sales, in addition to direct public support of R&D.

Open-source mechanisms provide a radical reorientation. They tend to directly promote innovation, rather than indirectly by fostering the financial interests of innovators. In various ways, they are geared toward supporting those initiatives in developing countries that are among the most promising in terms of sustainably alleviating poverty through the proper use of natural renewable resources. These initiatives, which do not depend on technological transfers under constraints imposed by developed countries, are often designed and implemented according to Einstein's rule: "As simple as possible, but no simpler."[1] For providing electricity in rural India, drinking water in Cambodia, healthy maize in Kenya, agroforestry protection in drylands, safe stoves wherever they are needed, innovators have in that spirit made the most of locally abundant and underutilized human and natural renewable resources.[2] Innovators have also trained local underemployed youth to fill technical and managerial positions in the enterprises they have created. And they have inserted their action within the surrounding social contexts, thus articulating natural and human capital in ways that are broader and far more efficient and useful than what would be achieved by concentrating only on technological achievements.

INTELLECTUAL PROPERTY RIGHTS: DR. JEKYLL AND MR. HYDE

STRIKING A REASONABLE BALANCE

A limited number of basic legal principles have traditionally regulated IPRs. To be patentable, an innovation should be new. It should not merely reproduce something that is already known; instead, it must entail an actual "inventive step." It should not be obvious. It should also have practical uses. According to this legal tradition, a discovery is not an invention and thus is not patentable; to distinguish between the two, innovations are humanmade, and discoveries are structured observations of natural phenomena. The breadth of a patent must correspond to the actual scope of the invention.

Economic analysis vindicates, to a large extent, the traditional legal principles mentioned here. There is no point in creating incentives to reinvent something that already exists, and there is a cost to making access to knowledge more restricted without commensurate benefits. When a patent is granted, there is a potential high cost in terms of privatizing knowledge, including the burden of all the transactions necessary to access that knowledge—hence, the necessity of strictly limiting the breadth of patents.[3]

Patents exist because they can provide a direct dynamic benefit in spurring innovation. They also have an indirect benefit in that other potential innovators are helped in their efforts by the information that must be disclosed when a patent is granted. That information would not be available if, in the absence of patents, innovations were kept secret. However, if its breadth is excessive, a patent will act more as a roadblock than as a stepping-stone to further innovations:

"When a broad patent is granted, its scope diminishes incentives for others to stay in the invention game, compared with a patent whose claims are trimmed more closely to the inventor's actual results" (Merges and Nelson 1990, 916). Economic analysis hence supports the traditional legal principle pertaining to the appropriate adjustment of a patent's breadth to the inventor's actual achievements.

Economists have produced more precise results by specifically investigating the problem of a patent's optimal breadth. A patent on an invention or a discovery should be the narrower (1) the fewer close substitutes there are for the products developed from the invention or the more difficult it is to bypass the invention or discovery in subsequent research; (2) the lower the cost of completing the invention; and (3) the higher the nonmonetary incentives (for example, "academic rewards") available to motivate the inventor.[4]

The last two conditions reflect the desirability of minimizing the effects of the deeply rooted imperfection (i.e., privatizing knowledge) associated with using patents as incentives to further innovation. The first condition implies that it is not appropriate to grant a broad patent to an invention or a discovery that, in turn, commands access to lines of research that cannot be pursued without the results covered by the patent. Under such circumstances, the invention or discovery is an "essential facility"—that is, it is essential for working on further research and invention. This marks the intersection of the economics of protecting intellectual property and the economics of protecting competition (including competition for innovation and the access to knowledge). As Tom and Newberg (1998), both members of the U.S. Federal Trade Commission at the time, put it: "If market power in an antitrust sense is not to be presumed, then, as with any other form of property, the existence of such power must be determined by evaluating the availability of close substitutes"[5] (346).

THE GLOBAL DEMISE OF THE BALANCE

There has been a remarkable increase in the number of patent requests submitted and accepted. Of course, the scientific and technical breakthroughs that occurred during the 1980s and 1990s contributed to that increase, but most of it can be attributed to a de facto

BOX 8.1 APPLYING THE PRINCIPLES: SHOULD GENES BE PATENTABLE?

Let's consider elements in the living world such as genes, proteins, and enzymes. Regarding such elements, which incidentally are discovered and not invented, even the caution urged by Merges and Nelson (1990) cited earlier might not be sufficient. From the viewpoint of economic efficiency, it might be necessary to reduce a patent's breadth further, below what might seem to be the inventor's marginal contribution to expanding the frontier of knowledge. In antitrust terms, these elements are "essential facilities"—they are necessary for others to do their own research. In addition, some argue that the process of isolating, sequencing, and characterizing genes and proteins has become routinized, with minimal costs.

For all these reasons—and the essential facility character is paramount—no broad patent should be granted on these elements, and possibly no patent should be granted at all. For instance, in the context of genetically modified food, Dietmar Harhoff, using the tools of industrial organization analysis, concluded that "granting patents on genes themselves (or even on gene functions) is not necessary to promote innovation. It is likely even to delay the development of socially useful applications" (Harhoff et al. 2001, 289).

On June 12, 2014, the U.S. Supreme Court started turning the tide of granting patents on genes, proteins, and other elements of the human body by ruling that raw genes—that is, genes in their natural condition—are not patentable.

reversion of the patent system to the role it had in the nineteenth century: essentially, a registration system.

Patents are now routinely granted to submissions devoid of any novelty or with insignificant original contributions. There are even cases of patents being granted to parties that are not the real innovators. Overlapping patents are granted, which is a sure recipe for igniting inextricable conflicts. Patents that are broader than they should be are the norm. The domain of what can be patented has been widely expanded, without a rational, balanced assessment of the benefits and costs in each case (see box 8.1). This is the case in applied mathematics, computing, and business methods, which have seen patents like Amazon's 1-click ordering and portfolio choice methods that boil down to the inversion of a matrix. Consider also the effort by a collaborator of Microsoft to get a patent on computer formulations of some Darwinian methods for testing the laws of evolution (Pennisi 2009). The patent offices' willingness to grant patents to traditional knowledge, such as basmati rice, neem juice (see chapter 1), and the healing properties of turmeric, has entrenched opposition to intellectual property rights in developing countries (Stiglitz 2006).

Dealing with patents, in or out of court, uses up as much or more effort and money than working on genuine innovations. As Robert Barr put it at a Federal Trade Commission Roundtable in 2002: "An innovator asks two questions: Can I get a patent? Do I infringe the patents of others? The answer to the first is usually too easy: yes. The answer to the second is much more difficult and, as a practical matter, impossible" (U.S. Federal Trade Commission 2003).[6] Barr could have added: if the innovator acts on the basis of a "yes" (or even a "maybe") to the second question, it may be very costly in time and money to disentangle the web of dependencies on existing patents. The patent system has made research an even riskier business: to the uncertainty about the research effort's success is added litigation risk. Small and medium-sized firms do not have enough resources to withstand a legal battle against large firms. In the current system, the small and medium-sized firms that could be particularly innovative are thus deterred from fulfilling their potential. The outcome is the worst of all possible worlds: not only is free access to knowledge reduced, but also the very function of patents—to act as incentives to innovation—is stifled by the proliferation of bad patents. Barr's is a businessman's assessment, but it has its parallel in science, as biologist David Maddison put it: "As patents enter this field, there is a very great danger that we will get bogged down in a legal morass" (quoted in Pennisi 2009, 664).

During the past thirty years, the United States has been active in exporting its unbalanced IPR regime around the world (see box 8.2). As a result, the country's dysfunctional approach to patenting (which involves granting too many patents and patents that are too broad) has spread globally.[7] To understand the forceful push of the United States in this direction, it is useful to remember the atmosphere of

BOX 8.2 HOW TO PERVERT A PATENT SYSTEM

Today's global IPR system has been greatly affected by the historical evolution of IPR in the United States.

In 1982, in an atmosphere of pessimism concerning the competitiveness of the U.S. economy, particularly in terms of trade in some high-tech sectors, Congress created the U.S. Court of Appeals for the Federal Circuit (CAFC) within the framework of the Federal Courts Improvement Act. The CAFC specializes in intellectual property matters, and it is the only U.S. court that handles appeals on such matters. The objectives in creating the court were to ensure greater consistency in dealing with appeals and to support an approach that would be systematically sympathetic to the defense and promotion of intellectual property. The judges chosen to sit on the CAFC were, to a large extent, selected according to their supposed willingness to further the latter objective. The statistics of the decisions made by the CAFC since its inception—including a dramatic increase in the number of rulings on patent infringements in favor of patent holders, as well as support to skyrocketing damages granted by the judges—reveal a pro-patent bias that is certainly not disappointing to the founders of the CAFC (Jaffe and Lerner 2004).

Congress has thus consciously promoted easier access to patenting. As controller of public receipts and expenses, it appears that it also unconsciously pushed in the same direction: by starving the U.S. Patent and Trademark Office of adequate funds, there is now a situation in which overloaded and underpaid examiners are not able to properly assess submissions for patents. As the system of incentives becomes more and more geared toward granting patents, it is only natural that examiners will tend to grant patents easily on the basis of generally superficial investigations.

technical and economic pessimism of the 1970s. The United States was seen as losing its competitive advantage in manufacturing, especially to Japan. At least U.S. universities and technological powerhouses were preeminent at the time. Because of Japan's expertise in adaptation, the United States believed it should get returns from its technological leadership. There was a general feeling in the United States that the absence of a proper global system of protection of IPRs was interfering with the country's ability to appropriate the returns to its investments in intellectual capital; thus, competition was seriously distorted. The idea emerged that the best remedy would be to introduce compulsory global rules on the protection of intellectual property into the mechanisms regulating free trade among nations.

For a long time, the World Intellectual Property Organization and its predecessors had worked to create global rules, but there was no enforcement mechanism. A practical way to have an enforcement mechanism was to link intellectual property with the trade agenda. Intellectual property thus became an item on the broad agenda of the Uruguay Round under the General Agreement on Tariffs and Trade (GATT).[8] Developed countries did indeed succeed in forcing the 1994 Trade-Related Aspects of Intellectual Property Rights (TRIPS) agreement on a reluctant developing world. The World Trade Organization (WTO) was created at the same time and was granted specific powers to arbitrate disputes and allow those hurt by unfair trade practices to impose sanctions on the offending parties. Among such unfair trade practices are listed violations as defined by TRIPS. The ultimate U.S. objective had been reached.

A crucial question is: Have the promises of substantial benefits for developing countries materialized? Are there more commercial investments from developed countries, as TRIPS allegedly makes such investments more secure? If there has been any increase due to TRIPS in the flow of investment, it has been relatively modest and mainly by subsidiaries of multinationals. In what might at first glance appear as a paradox, the largest flow of commercial investments by far has been going to China, the country that has most consistently been accused of cheating on TRIPS, even after it formally endorsed the agreement when it became a WTO member.

Even from the outset, it was recognized that the TRIPS agreement was unbalanced, with costs imposed on developing countries manifestly greater than the benefits. For developing countries, the globalization of intellectual property has had two main consequences: increasing havoc in their public health systems and draining royalties toward rich countries (Newfarmer 2001). This global state of affairs is obviously not geared toward disseminating innovations that might put developing countries on sustainable development trajectories. What then are possible changes to remedy this situation?

DUAL ENGINES: REGULATION AND COMPETITION

HAVING THE RELEVANT INFORMATION REVEALED

A key problem noted earlier is the granting of patents that should not have been granted. Not only does this result in excessive "privatization" of knowledge; it also leads to excessive litigation costs, and, as we have seen, it stymies innovation.

The European Patent Office (EPO) has a procedure for evaluating the validity of patents that seems preferable to those employed elsewhere: when a patent is granted, parties that are unhappy with the decision and that think they have robust arguments to prove the patent is unwarranted may demand an "opposition" procedure before an appellate body within the EPO. Such a procedure is quicker and far less costly than going to court. Most important, the appellate body considers all significant evidence that is submitted. The opposition procedure functions as a device that elicits and examines relevant information that the opposing parties possess and have every interest to communicate. This is a particularly important function in situations in which the quality of direct information gathered by the examiners in patent offices has seriously deteriorated, partly due to budget constraints that lead to lack of experienced staff.

This function of the opposition procedures is so important that Nobel laureate in economics Jean Tirole (2003) suggested that it should be integrated within the examination process itself. Jaffe and Lerner (2004) concurred when making the following recommendation: "Create incentives and opportunities for parties that have information about the novelty of inventions to bring that information to the PTO when it is considering a patent grant." Also in agreement was Robert C. Pozen, who wrote an editorial in the *New York Times* (November 16, 2009) to reform the way patents are examined:

> Patent examiners, many of whom are young or lack practical experience, are not qualified to evaluate whether complex claims in biotech or physics meet the most critical tests: whether the claim is novel relative to prior art, and whether this would be obvious to a

> person skilled in the art. To help fix this, Congress should pass an amendment allowing experts in the field to submit explanatory or critical comments on patent applications.[9]

Such reforms would drastically reduce the mountain of bad patents that are routinely granted, as well as the excessive breadth of many patents. Opposition procedures would be a good example of a revelation mechanism within which the parties involved have strong incentives to reveal the information they possess—information that is of paramount importance for reaching an appropriate decision.[10]

COMPETITION AS INCENTIVE

Traditionally, advocates of intellectual property have argued that the economic distortion associated with the underutilization of knowledge—and the potential reduction in competition—is more than offset by the benefits of greater innovation. More recently, however, this perspective has come under two criticisms. First, innovation itself is hampered (see previous discussion). Second, there are better ways of providing incentives for innovation without the adverse effects associated with the patent system.

John Barton chaired a commission appointed by the UK Department for International Development (DFID) with the objective of integrating intellectual property rights and development policy (Commission on Intellectual Property Rights 2002).[11] The Commission on Intellectual Property Rights devoted a great deal of attention to health and agriculture issues, pointing out serious difficulties in integrating scientific and technical innovations into development policy due to the monopolization of genes and other elementary constituents of life by current intellectual property law and practices. Five years later, Barton completed another report on intellectual property and development in the field of clean technologies in which he identified no roadblock similar to genes, with the possible exception of enzymes for the production of biofuels—at least photons and electrons have not yet been considered for patenting (Barton 2007). Moreover, he stated that "there is competition between a number of patented products"—that is, between techniques and devices to produce clean energy.

Such competition may be promoted at all levels, including the process through which patents are granted, which currently only grants gold medals. Why not grant silver medals as well, so that inventors applying within a sufficiently short period after the winner of the race (a claim may take years to be examined) would share the patent? One strong objection to the idea of "silver medals" is that it would weaken the incentives to invent. However, that is not necessarily the case; the chance of receiving several lesser prizes might compensate for the absence of a jackpot, especially when, as is often the case, there is some degree of differentiation among the proposed inventions (E. Henry 2010).

In the "silver medal" reform just described, patents do get issued—albeit, shared—in ways that better balance incentives to invent and benefits from competition. As James Bessen and Eric Maskin (2009) explained, if the proper balance of certain economic activities and structures were achieved, they could dispense with patents altogether. Bessen and Maskin observe that "such industries as software, semiconductors, and computers have been so innovative despite historically weak patent protection." And that is because "when innovation is sequential and complementary, standard conclusions about patents and imitation may get turned on their heads. Imitation becomes a spur to innovation, whereas strong patents become an impediment."

In this case, there are "natural market forces" that call for innovations and that dynamically protect innovators from imitators.[12]

OPEN SOURCES OF INNOVATION

The open-source movement has deeply influenced the production of software over the past two decades. The "success of open source," to use the title of Steven Weber's book (published in 2005), is manifest. Linux, one of the most prominent open-source projects, is dominant in web servers and has a significant presence on other markets. Apache, another open-source web server software, now serves more than 50 percent of all websites. However, the success is not limited to a few prominent examples. SourceForge, the largest open-source software development website that provides tools and services for developers, is currently hosting more than 250,000 projects and has millions of registered users. Although many of these projects are small, these numbers reflect the vibrancy of the movement.

Far from reveling in anarchy, the open-source model relies heavily on contracts. A piece of software is generally licensed under a "copyleft" license, the most prominent one being the General Public License (GPL). The idea is that a licensee can freely use the software for any purpose. The central restriction placed on the licensee is that modified versions also need to be licensed under copyleft licenses. Firms can make profits in this environment, distributing precompiled versions of the software, providing assistance to users (among them are many large companies), and devising specific features when required.

Right now, only the software industry has seen the penetration of open-source contracts on a large scale. However, there are significant examples in other sectors.[13] The case of the open-source contracts on the bacteria identified by the Cambia scientists in Australia (see chapter 6) provides an important example in biotechnologies in which the innovations may have remarkable potential in terms of sustainable development—in particular, poverty alleviation.

A WORLD OF INITIATIVES

Open-source contracts represent a radical reorientation in intellectual property protection, from an approach permeated by the objective of restricting access to an approach that aims to keep access free and to strictly protect this freedom. This latter approach is in tune with the necessity of having a large array of unimpeded initiatives toward more sustainable modes of development: "Technological solutions to global problems must be deployed throughout the world by many different actors in decentralized ways," in Richard Nelson's words (Mowery, Nelson, and Martin 2010, 1012).

Let us now investigate how to create and support decentralized initiatives for designing and deploying systems—with their various natural, technical, social, and economical components—meant to sustainably meet fundamental needs in developing countries.

We base our discussion on some remarkable endeavors aimed at providing electricity, drinking water, or food to disadvantaged communities; we also consider a worldwide effort to improve the health and environment of about two billion people who rely on primitive stoves for cooking and heating. The following factors prove critical for sustaining innovation and development in such circumstances:

- Entrepreneurship
- Technical versatility, in the spirit of Einstein's rule
- Local underutilized human and natural renewable resources as main inputs in the production processes to save natural capital and promote human capital
- Innovative education and knowledge dissemination

HUSK POWER SYSTEMS: PROVIDING ELECTRICITY IN RURAL INDIA

In 2007, Gyanesh Pandey, an Indian engineer who had graduated in electrical engineering from Rensselaer Polytechnic (Troy, New York) and who had a good job in Los Angeles, decided to head back to his native Bihar, India. Bihar is mostly rural and is one of the poorest states in India. More than 80 percent of households have no access to electricity, a proportion that both reveals and breeds poverty. Those who can afford them use inconvenient and costly kerosene lamps that generate indoor pollution; diesel generators, also polluting and costly, are used to pump water for irrigation and sustain artisanal and commercial activities.

Pandey did not come from a well-off family; as a child, he suffered from lack of proper lighting. By 2007, he was determined to muster his technical skills to remedy the situation in his home state. After a few unconvincing attempts with solar cells and biofuels, he came to the idea of using rice husk to generate electricity. He teamed with a local entrepreneur and two Indian graduates from University of Virginia Darden School of Business. Husk Power Systems was started in 2009.

In Bihar, rice is the dominant crop; husk—that is, the envelope of the rice grains—is thus abundant. Husk is good neither for burning in stoves (because of its very high silica content) nor for returning nutrients to the soil (because of its low nutrient content). However, it can be decomposed by fermentation in a gasifier. Because it had very few uses, 75–80 percent of the two million tons obtained each year as a byproduct of the rice crop was rotting in landfills. The resource was thus plentiful, and its use as a precursor of fuel didn't harm any other activity.

At Husk Power Systems, small simple gasifiers are fed with husk. The gas is then burned to drive a turbine from which electricity is produced in a standard way. Typically, a 30–40 kW power plant consumes 50 kg of husk per hour. The components from which these mini power plants are made are not tailor-made; they are bought in such conditions that costs are minimized. However, their arrangement into specific equipment is innovative, with a quest for simplicity and efficiency in using an unusual fuel.

Typically, the investment cost is about $1,300/kW, which is partially paid for by consumers and partially by modest grants from the Indian federal government, the International Finance Corporation, and foundations like the Shell Foundation and the Alstom Foundation. The variable cost is about $0.15/kWh, which is covered by consumers in counterpart for the delivery of enough electricity for one or two low-consumption bulbs and mobile phone recharges. What consumers pay is about half the expense they would pay to use a kerosene lamp. Electricity is distributed through local mini networks—that is, simple wiring of a few villages totaling up to 300–500 households. The electricity makes it possible for residents to extend their home activities—in particular, student work—beyond daylight hours. For artisanal and commercial activities, they enjoy a less polluting, more convenient, cheaper source of energy, making their businesses more productive. Each local network allows the saving of about 40,000 L of kerosene and 18,000 L of diesel per year.

At the Husk Power Systems "university" (in German, it would more accurately be called "Technische Schule"), most students are recruited locally. They are trained either as a "plant's junior mechanic," with the goal of being put in charge of the operation and maintenance of a single plant (an eight-week course), or as "senior mechanic" and middle manager for a number of plants, with the ability to face more intricate problems than those dealt with at the plant level (a six-month course).

Thus, Husk Power Systems is more than a technical innovation, however valuable it is in that respect. It integrates into the economic life of the communities a local, abundant, and underutilized raw resource. It also promotes local talents. By providing an essential service, it transforms the economic, social, and health conditions of the communities it serves.

Within the past four years, about eighty plants and networks have been established, with cumulative improvements in service and costs. The pace of development is accelerating, and inroads have been made into the neighboring state of Uttar Pradesh, as well as East Africa; there is also interest in Bangladesh (Islam and Ahiduzzaman 2013). Pandey concludes that the main lesson of the endeavor is how to create a system providing an essential service, adapted to the needs of poor people, out of the material and labor resources that are readily available locally.

1001 FOUNTAINS: PROVIDING DRINKING WATER IN CAMBODIA

The nonprofit 1001fontaines (1001 fountains) aims to provide safe drinking water in rural Cambodia. Although it was designed independently from Husk Power Systems, their structures have striking similarities that reflect converging assessments of similar circumstances. In both cases, the following are true:

- Entrepreneurship is a driving force (Rambicur and Jaquenoud 2013).
- Technology is "as simple as possible, but no simpler," with special attention to the system's reliability and the product's safety.
- Production is centered on a preexisting community of consumers, such as a village or a small cluster of villages.
- There is a duality between (1) local plants run by local technicians-managers recruited from the local communities and trained in the organization's "academy" (similar to the Husk Power Systems "university") and (2) a central body that handles problems that cannot be solved at the local level and that sustains the geographic expansion of the system (60 plants in 2011, 120 in 2013, and 250 in 2015–2016).

- Interactions with consumers are closely monitored. In the case of 1001fontaines, the priority is to convince villagers that getting safe drinking water is worth paying a price that, however modest, is not negligible for them. Husk Power Systems has similar concerns.

PUSH-PULL SYSTEMS: PROTECTING CROPS IN KENYA

A push-pull system is a biology-based method of protecting crops. In the example here, this system is used to protect maize in East Africa (the region's main crop). It is the product of a joint project at the International Centre of Insect Physiology and Ecology (ICIPE, Kenya) and Rothamsted Research (UK), one of the longest running and most productive agricultural research stations in the world.[14] The targets of the system are maize stem borers—the larvae of various moths that attack maize from within the stems. If left unchecked, stem borers can reduce yields by 20–40 percent and sometimes up to 80 percent. As research into the system progressed, a second target popped up by chance—or, more accurately, by recognition of an unexpected side effect. *Striga hermonthica* is a weed that is extremely difficult to eradicate by conventional methods, as it parasitizes the maize; the yield losses range from 30 to 100 percent in infested fields.

Pesticides are not very effective at reaching larvae inside stems. Thus, an all too common reaction is for farmers to increase the quantities applied, which results in more harm to soil biodiversity than to pests. Yet, the stakes are high; according to ICIPE (2015), "Preventing crop losses from stemborers could increase maize harvests by enough to feed an additional 27 million people in the region" (3). This was a strong motivation to try to find effective ways to at least keep the stem borer populations within limits so they cause little harm, while minimizing negative effects to soil and environment.

Researchers considered a few candidates for the push and a stunning 400 varieties of grass for the pull. Their research led to identification of a tree (a tree that also fixes nitrogen from the air as a bonus), called silverleaf desmodium, for the push, and two varieties of grass (Napier grass and Sudan grass) for the pull. These plants were selected for several interrelated reasons:

- When intercropped with maize, desmodium repels female moths and thus deters them from laying their eggs on the maize lines. That worked as anticipated; what was not anticipated was that the desmodium roots emit into the soil a chemical substance that checks the growth of the *Striga* weed. In the experimental fields with maize and desmodium, the progressive disappearance of *Striga* was observed with a measure of surprise—and then great satisfaction.
- Napier and Sudan grasses, which are planted along the borders of maize-desmodium parcels, emit volatile chemical substances that attract female moths and make them lay their eggs there, instead of in the maize.[15] The grass is also home to various predators of the eggs and of the larvae coming out these eggs, including ants, earwigs, spiders, and a remarkable variety of tiny wasps that parasitize eggs. Another reason for choosing these grasses is that they provide valuable fodder for livestock.

Adoption by 50,000 smallholders as early as 2012 has resulted from associating farmers to the experiments, from broad knowledge dissemination and demonstration of results: "The solution was to recruit some of the more experienced farmers as teachers to help their colleagues" (8).

Adoption has taken place at an accelerating rate, which makes a million adopters by 2020 a realistic objective. This rapid adoption is mainly the effect of demonstrated increases in yields: control yields on maize monocrop fields are typically between 1 and 2 tons/ha/yr; on push-pull fields, they are between 4 and 5 tons/ha/yr, with less volatility.

At this point, Voltaire would warn against the Candide delusion. Indeed, with all its virtues, push-pull is not without its problems: the broad use of Napier grass causes the spread of a disease hitherto marginal, and desmodium is attacked by its own variety of bore. Nothing is more inventive than life, with positive effects that human endeavors can take advantage of and negative ones that we must try to circumvent. In this case, researchers are back to work, trying inter alia to identify and transfer resistance genes among various strains of desmodium and Napier grass.

AGROFORESTRY AS A GREAT GREEN WALL OF NATURAL CAPITAL-ENHANCING VENTURES

The idea of planting an east-west "Great Green Wall" in Africa in order to contain the expansion of the Sahara desert and to stem ubiquitous land degradation was formulated in the 1950s by British scientist Richard St. Barbe Baker. The idea was revived in 2005 by Olusegun Obasanjo, then president of Nigeria, and was endorsed by the Assembly of the African Union. The "Great Green Wall Initiative" was officially born; however, it soon evolved from the initial vision of planting a transcontinental tree belt to a common resolve of supporting decentralized green ventures comprising not only tree windbreaks but also developments in agroforestry.

In southern Tunisia, a region damaged by advances of the desert and extended periods of drought, agroforestry development has been pioneered by Sarah Toumi, who won the nickname "La Dame aux Acacias."[16] She started by convincing local farmers—women, in particular—to plant acacias in their fields to form grids among the usual crops. A great many varieties of acacias can be found in sub-Saharan Africa and Australia. Some of them, for instance *Faidherbia albida*, are landmarks in the African savanna and are not only beautiful but also astonishingly useful in many respects. First, they are legumes: with the cooperation of specific bacteria (see chapter 2), they fix nitrogen from the atmosphere, thus fertilizing the soil for their own benefit and the benefit of surrounding crops. In sub-Saharan Africa, up to 300 percent increases in maize yields are recorded in crops interspersed with acacia trees; large increases have also been observed in millet and sorghum fields planted the same way. Second, they are vegetables: their leaves are eaten in salads and other dishes and are even stored as nutritious powder. From their seeds, edible oil is extracted. Acacias also produce Arabica gum, an input in various industrial processes. Third, their extended root systems are highly efficient soil stabilizers. Fourth, they improve water circulation and storage, provided the grids in which they are organized are neither too dense nor too loose. If too loose, they have no significant effect; if too dense, the trees consume too much water by transpiration and evaporation on the leaves. With properly scaled grids, however, positive effects are dominant: the tree cover reduces evaporation from the crops underneath, and the roots infiltrate water deep into the soil where it is stored, to be recovered later in the dry season, thus reducing the volume and erosive power of surface runoff (Ilsted et al. 2016). Fifth, the trees provide wood suitable for making furniture and for incorporating into buildings. Indeed, according to Daniel Hillel, "Acacia was chosen to provide wood for the Tabernacle" during the exodus (Hillel 2006, 297).

For local populations, these are most valuable benefits. Take, for example, the leaves. Because the acacia is drought-resistant, losing its leaves during the wet season and growing them back during the dry season, the leaves are available when other crops are not. Being able to rely on acacia leaves will become a matter of food security as drought conditions get more prevalent due to climate change and desert encroachment.

Toumi's effort has resulted in tens of thousands of acacia trees having been planted. She has since suggested introducing diversification with *Moringa oleifera*, a tree originating from India and present in some parts of sub-Saharan Africa, though unknown in Tunisia. Much smaller than *Faidherbia albida*, the moringa represents an intermediate level in an agroforestry setting. With its phenomenal nutritional potential, it was elected as "traditional crop of the month" by the Food and Agriculture Organization on the basis of the following assessment:

> All parts of the moringa tree—bark, pods, leaves, nuts, seeds, tubers, roots, and flowers—are edible. The leaves are used fresh or dried and ground into powder. The seed pods are picked while still green and eaten fresh or cooked. Moringa seed oil is sweet, non-sticking, non-drying and resists rancidity, while the cake from seed is used to purify drinking water. The seeds can also be eaten green, roasted, powdered and steeped for tea or used in curries.[17]

In a tribute to traditional local uses, the FAO report praises moringa leaves for having more beta carotene than carrots, more protein than peas, more vitamin C than oranges, more calcium than milk, more potassium than bananas, and more iron than spinach. Similar to the *Faidherbia albida*, the leaves come in the dry season, making them an outstanding contribution to food safety and dietary balance.

Furthermore, the same presentation by the FAO mentioned that "moringa products have antibiotic, antitrypanosomal, hypotensive, antispasmodic, antiulcer, antiinflammatory, hypocholesterolemic, and hypoglycemic properties." It comes as no surprise that moringa products are finding their way into the health stores of developed countries. For the farmers in Tunisia, this migration is a mixed blessing: it might bring them some added cash, but it also makes the moringa products less available for meeting their own needs (see chapter 1 about the U.S. holdup on the fruits of the Indian neem tree).

Moringa's soft wood is of limited value. However, because it grows at an unusually fast pace (and regrows after cuttings), it can provide large amounts of firewood, sparing women and girls long distances of wood fetching and sparing forests of destructive cuttings. Combining firewood sourced from agroforestry with relatively efficient cooking stoves (see next section) cures several ills at once.

Agroforestry projects, like Toumi's work with acacia trees in Tunisia, have been around for millennia, in many places and under various guises, though it is of greatest value in drylands. The empirical knowledge that has accumulated along the way is highly valuable. However, mere empirical knowledge is no longer sufficient under shifting climatic and demographic conditions. Moreover agroforestry should be made available to the hundreds of millions farmers who would benefit from adopting suitable variants of it, even for those who live in places with no previous agroforestry experience. Here is a typical case: Mr. Maskuri is a farmer in Konawe Selatan in Indonesia's Southeast Salawesi province. Speaking for his neighbors and for himself, he told researchers from the World Agroforestry Centre:

> Farmers in my village didn't know much about good agroforestry practice until last year. All we knew about growing a plant was to put it in the ground and hope it would live so we could make a living out of it. And, of course, the plants died! I did that for years since I moved here in 2008, but it was only in 2014 that I finally managed to successfully plant cocoa for the first time, after

> I got help from the AgFor team. Now, our farmers' group plans to implement agroforestry in all our gardens, such as combining rubber and orange, cocoa and coconut, or jackfruit and pepper. (Lumban Gaul 2015)

New concerns must also be taken into account—in particular, the capacity of both vegetation and soil to store carbon. Trees are good at that, but some are better than others (Toensmeier 2016; the author also reviews a stunning variety of plant combinations suitable for agroforestry systems)—for example, acacia is good, but it is not among the best. Farming methods also matter; Johan Rockström, for example, stressed that "one of the most promising practices is abandoning the plough, which has proven to be an effective way of burning soil carbon—by exposing it to sun and oxygen—causing emissions of carbon dioxide" (Rockström and Falkenmark 2015, 285).

To tackle pressing challenges in order to explore, implement, and disseminate new ecological and technical approaches, it is imperative to make better use of human resources. In this respect, it is crucial to bring women and young people to the fore. Because they lack opportunities in most traditional communities, they welcome innovations that unsettle the established order. "If you pay no attention to gender then you set up to fail," said Catherine Bertini (2014), a former executive director of the UN World Food Programme during her Sir John Crawford Memorial Address on investing in women. Sarah Toumi has made every effort to help create associations in which projects, achievements, and difficulties are freely discussed and in which contributions by women and young people are encouraged and valued within a collective dynamic.

Modernizing agroforestry in the sense of designing and promoting appropriate configurations of plants and efficient working practices, adapted to local ecological and human conditions, is essential. This task is unfortunately made unduly difficult by the world dominance of monoculture-oriented, chemistry-based agriculture. Money for necessary investments is not made readily available by financial institutions, whether public or private. Specific R&D is not properly organized and supported, especially for breeding, which could improve yields of traditional crops by wide margins. These deficiencies are all the more difficult to overcome because agroforestry systems comprise a number of diverse components, which is at the root of their aptitude to provide broad arrays of products and services and to generally avoid uncontrolled proliferation of pests. However, it also puts them in an awkward position in relation to narrow-focused institutions with routines that are at odds with the versatility of agroforestry. In these respects, Sarah Toumi's dealings with bureaucracies rarely boosted her endeavors.

Where farmers have only weak rights on the land they farm, they are under permanent threat of land grabbing (see chapter 2); hence, they are deterred from making the investments required to adopt and maintain agroforestry. Lack of long-term tenures is thus another serious obstacle against the expansion of agroforestry. Tenures are crucial entitlements, the importance of which has been stressed by Amartya Sen (1983): "A person's ability to command food depends on the entitlement relations that govern possession and use in society" (154).

Summing up, we have seen that agroforestry regenerates soils and regulates water flows; thus, it reverses land degradation and helps overcome droughts. It also enhances carbon storage. It makes fundamental contributions to food safety and dietary balance. According to a report by the United Nations Environment Programme Economics of Land Degradation (ELD Initiative and UNEP 2015), "The benefits of taking action in Africa are almost 7 times the cost of action" (11). The list of new agroforestry systems being settled is long—despite the obstacles mentioned—not only in Africa, but also in Asia and Latin America. In Indonesia, it is seen by farmers as the only way to face ever more severe droughts. In Thailand and Vietnam, it aims at balancing the expansion of *Hevea* monoculture fueled by the Chinese demand for rubber. In Malawi and Zambia, the priority is to replenish soils depleted of nitrogen. In North Africa, it is to stem desertification. In Cameroon, it is to sustain pioneering efforts started twenty years ago. And so on. However, dissemination of information remains insufficient. Properly focused R&D, as promoted by the World Agroforestry Center and as embodied in Sarah Toumi's partnership with the Sahara and Sahel Observatory, would bring improvements commensurate with its present underdevelopment. Local and regional leadership is paramount to overcoming psychological, sociological, and institutional barriers. On both counts, "Acacias pour Tous" ("Acacias for All"), Sarah Toumi's flagship, has much to show.

CLEANER FUELS FOR IMPROVED STOVES

Some endeavors are international by construction, not merely by extension from a regional or national basis. The Global Alliance for Clean Cookstoves (GACC), a public-private endeavor hosted by the UN Foundation, was created in 2010 with the objective of improving health and environment conditions for people in developing countries who currently cook and heat their houses with primitive stoves or open fires. GACC supports a great variety of decentralized efforts to design and produce safe and efficient stoves, and the corresponding fuels, that are adapted to local circumstances (Global Alliance for Clean Cookstoves 2015).

As a contribution to the "Global Burden of Disease Study," the *Lancet* (2012) published "A Comparative Risk Assessment of Disease and Injury Attributable to 67 Risk Factors." The top three factors were high blood pressure; tobacco smoking, including second-hand smoke; and household air pollution from solid fuels. Indeed, according to the International Energy Agency (2006): "In developing countries, especially in rural areas, 2.5 billion people rely on biomass, such as firewood, charcoal, agricultural waste and animal dung, to meet their energy needs for cooking. In many countries, these resources account for over 90 percent of household energy consumption" (419). And the side effects are devastating:

- Millions of premature deaths each year and tens of millions of debilitating ailments, with the victims mainly women and children.
- Wasted time and energy in collecting and transporting fuelwood, often at distances of 5–10 km, again involving mainly women and children. Collecting fuelwood is hard work and, in certain regions, puts women at risk of aggressions. Moreover, it precludes opportunities of education or more productive activities.
- Emissions of GHGs (CO_2, methane, nitrogen compounds) and black carbon, which could be dramatically reduced with more

efficient burning practices. In particular, black carbon, a volatile byproduct of incomplete combustion, is emitted on a large scale by household stoves, which account for about 25 percent of black carbon from all sources worldwide. It is a powerful absorber of direct solar radiation wherever it settles, particularly on ice (as in the Eastern Himalayan glaciers, as discussed in chapter 4).
- Severe degradation of local forest ecosystems.

The consequences for people and for natural capital are obviously harsh.

Several types of stoves, along with the corresponding fuels, are now on offer that are vastly more energy efficient and cleaner than the traditional devices (International Energy Agency 2006; Reddy 2012). The more affordable ones, which are also closer to the cooking habits of the households concerned, are improved biomass-fired stoves. The fuels are conditioned and the stoves designed in ways that significantly reduce the side effects; in particular, they minimize black carbon emissions and instead produce a stable solid carbon residue called biochar, which contributes to carbon sequestration and can be used to improve soil fertility.[18] More sophisticated systems function with biogas, which is the best choice when available (Surendra et al. 2014) and when electricity is not. Others function with liquid fuels, solar heat, or electricity.

A broad offer of efficient stove-fuel systems is a precondition for adoption, but it is not a sufficient one. Several other factors weigh on household decisions, such as the following:

- Household incomes, taking into account extra income generation allowed by better stoves
- Prices of stoves and fuels, adjusted for public or private financial support as needed
- Quality of stoves—in particular, their reliability (a condition for allowing the cook to save time for other activities) and their compatibility with cooking traditional meals (e.g., traditional bread in India, all sorts of tortillas in Mexico, and millet-sorghum galettes in Western African countries)
- Integration in the local economy, as local production of stoves increases their appeal to consumers

By and large, the interplay of these factors currently doesn't lead to as broad an adoption as one would like, though it seems quite possible that the GACC's objective set for 2020 (100 million new clean stoves replacing traditional ones) might be reached. Such an objective, however, pales in respect to the far higher number of households in need of replacement. A much bigger effort is required that should be adequately informed by the buildup of knowledge about how households make their decisions (Malla and Timilsina 2014). There could not be a more valuable contribution to sustainable alleviation of poverty, by protecting health, freeing people from exacting tasks, and conserving natural capital essential to communities.

9

ECONOMIC INSTRUMENTS FOR SUSTAINABLE DEVELOPMENT

ANNOUNCING THE introduction he was writing to the 250th anniversary edition of Adam Smith's *The Theory of Moral Sentiments*, Amartya Sen wrote in the *Financial Times* ("Adam Smith's Market Never Stood Alone," March 19, 2009):

> It is often overlooked that Adam Smith did not take the pure market mechanism to be a free-standing performer of excellence, nor did he take the profit motive to be all that is needed. . . . What is needed above all is a clear-headed appreciation of how different institutions work, along with an understanding of how a variety of organizations—from the market to the institutions of state—can together contribute to producing a more decent economic world.[1]

This quote defines the spirit of this chapter: we must make the most of market mechanisms and market-compatible mechanisms within a framework of rules and institutions caring for aims that a "free-standing" market cannot but ignore.

The chapter begins by addressing how efficient pricing of natural capital components may emerge from public finance reform or from the creation of dedicated markets. Both Sweden and the United States have successfully explored these two approaches: Sweden, with its sweeping fiscal reform of 1990, and the United States, with its Clean Air Act Amendment of 1990, which established a national market for sulfur dioxide (SO_2) emission permits. In contrast, the tentative agreement at Kyoto to establish a broad international market for CO_2 emission permits has failed, in part because some essential conditions of the U.S. success were not met on the global scale. Apparently, Adam Smith's teachings have been ignored. Sweden, for its part, has accelerated its transition to a sustainable economy and society, although its followers along this path are far too few.

The so-called divest (away from high-carbon footprint assets)–invest (toward assets in harmony with a decarbonized economy) movement on financial markets has gained momentum at a pace that was completely unexpected as late as 2014. When investors have rebalanced a portfolio in this way, they are motivated to support those public policies that enhance the value of the new investments—for example, stricter pricing policies. In this sense, private interests and natural capital–friendly public policies might come to converge. Although this would be a positive outcome, no meaningful convergence is possible as long as massive public subsidies and private investments continue to support the production and consumption of fossil fuels, not to mention, more generally, the destruction of natural capital such as oceans, forests, rivers, and wetlands.

PRICING: TAXES AND MARKETS

SWEDEN PIONEERS ENVIRONMENTAL TAXES

The first UN Conference on the Human Environment took place in 1972, at the famous Stockholm City Hall. Each participating country—there were only thirty-seven at that time—had been asked to prepare a case study of particular relevance to its own environmental context. The Swedish case study was centered on acid rain, which was then of great concern to Sweden. The study provided an assessment of the damages and causes of acid rains. It also served as an inquiry into possible remedies, with special reference to the potential role of economic incentives insofar as they might induce changes in the behavior of economic agents whose emissions of SO_2 and nitrogen oxides (NO_x) were prompting the acid rain. Among the economic instruments susceptible to generating the required incentives, particular emphasis was laid on taxes on emissions as a way of efficiently pricing the damages.[2]

Almost twenty years passed before the Swedish parliament (the Riksdag) brought this approach into law in 1990. As late as 1986, most Swedish people and politicians still believed in a command-and-control approach, blended with appeals to civic-mindedness. Putting a price on polluting emissions was largely seen as selling nature to those rich enough to buy it. Meanwhile, however, forests were decaying, and fishes were dying in acidified lakes. It is no surprise that these issues proved critical during the campaign for the 1988 general parliamentary elections.

Changes in the public mood came from an unexpected corner: in an astonishing U-turn, the small—but on the rise—Green Party proposed a grand bargain. As the traditional approach clearly wasn't working, they proposed a comprehensive set of "green" taxes at levels high enough to sufficiently deter polluters. Their logic was that pollution, however reduced, would never be completely eliminated and that any additional revenue from the tax would accrue to the Treasury. According to the proposal, the additional revenue should be exactly compensated by reductions in current taxes, income taxes, and corporate taxes, the high levels of which were hindering the Swedish economy and overburdening Swedish citizens.[3] The Social Democrats, the dominant party in Sweden at the time, jumped on the train and campaigned alongside the Greens. Together they secured a majority in the Riksdag and, in 1990, passed a law of fiscal reform along the lines promoted during the campaign.

The law targets most polluting activities. Emissions responsible for air pollution are taxed at rates unheard of elsewhere: around €3,000 (approximately US$4,000) per ton of SO_2 and €450 per ton of NO_x emitted from stationary sources (those that produce at least 50 MWh of useful energy per year), with the objective of eliminating 75 percent of SO_2 and 50 percent of NO_x pollution. It took Sweden ten years to reach the SO_2 target and seventeen years to reach the NO_x target.[4] The law also deals with water and soil pollution—for instance, by taxing chemical inputs in agriculture that are detrimental to water, soil, and the environment in general—and with garbage disposal. Anticipating a major environmental threat (i.e., climate change), the law imposes a tax of €27 (€7 for exposed industries) per emitted ton of CO_2. That tax has since been increased to levels that are seen elsewhere as stunning; the levels in the 1990 law were definitely not meant to be final destinations.

EMISSION PERMITS IN THE UNITED STATES: AN EFFICIENT MARKET

The United States also saw its share of success. In 1990, the U.S. Congress passed an amendment to the Clean Air Act that embodies the major endeavor to enlist economic instruments against a specific external effect—that is, air pollution from SO_2 emitted by power plants. That both the Swedish and U.S. initiatives became laws the same year is a coincidence, aside from the fact that both countries were severely suffering from SO_2 pollution and both were in need of fresh approaches.

During the 1980s, it had become clear that the command-and-control system—a Soviet-like approach, as it was sometimes derided—was unable to check increasing air pollution in the United States. In addition, the system imposed costs on firms in such an inefficient and incoherent way as to make two polluters in the same city face marginal abatement costs that differed by a factor greater than twenty. Despite opposition from beneficiaries of such discrepancies, President George H. W. Bush took advantage of a favorable political conjunction within Congress to reach an agreement centered on a system of economic incentives, with objectives regarding the reduction of SO_2 pollution comparable to those set in the Swedish law. Although the same incentives are at work in both cases, they are in different institutional settings and with different distributive effects. Rather than going the fiscal way ("no new taxes"), the amendment created a new market in emission permits (also called *allowances*); this new market was in rights as immaterial as those traded on financial markets.

The U.S. law targets power plants, as, at the time, a majority of them were burning coal that often contained high amounts of sulfur.

Every sizable power plant in the country got an initial allocation of permits recorded in an annex to the law, the result of hard-fought compromises. For every ton of SO_2 emitted during a set year, a plant must produce one permit valid for that year. If the plant's total emissions exceed its allocation of permits valid for the year, it must buy the missing permits on the market; if its total emissions don't use up its allocation, it may either keep the surplus of permits for later use or sell it on the market. A plant may also choose to reduce its emissions by shifting to fuels with lower sulfur content (for instance, coal from Wyoming instead of coal from the Appalachian Mountains), by filtering emissions, or by reducing output.

The market for permits has the effect that all such actions—including, of course, permit buying and selling—are available to all plants and thus tend to equalize their marginal costs of reducing pollution. This is the efficiency objective that was sought by creation of the market. The public good objective—that is, the reduction of total SO_2 pollution—is mandated in the law: the total number of allocated permits caps the pollution at a fraction of the actual total pollution in 1990. The reduction effectively achieved is somewhat smaller than the reduction achieved in Sweden, but the difference is not considerable. The market for permits displays a high degree of flexibility. Everybody may buy permits—for instance, the American Lung Foundation bought permits in order to tighten the cap mandated by the law. And it is not merely a spot market: transactions on permits for future years are possible at any time.

However, it is not the market free from all public interference that some senators had congratulated themselves on having launched. Indeed, it has succeeded not only because of its flexibility and fluidity, but also because it works within a well-structured and well-regulated framework. The regulator in this case is the EPA, which registers transactions, making sure there is no fraud. The EPA is entitled to buy and auction permits to prevent excessive tensions and accidental volatility on the market. It is also endowed with the duty, and the corresponding powers, of preventing anticompetitive dealings on the market, such as attempts at manipulating the price of permits. All of this is in line with the amendment's general philosophy: to make broad use of efficient market mechanisms within a framework of rules properly defined and rigorously implemented.

THE FAILED AMBITION OF CAPPING GLOBAL CO_2 EMISSIONS

In the fall of 1997, delegates from almost all member countries of the UN were converging to Kyoto, Japan, where a Conference of the Parties (COP) to the 1992 United Nations Framework Convention on Climate Change (UNFCCC) was about to start. The delegates were expected to agree on a protocol introducing mandatory reductions in CO_2 emissions from developed countries (labeled Annex B countries), while developing countries would only be invited to make efforts on a voluntary basis toward containing their emissions. This differentiated treatment was intended to reflect differences not only in development needs but also in responsibilities for past emissions.

Every Annex B country was assigned a specific rate of reduction, with 1990 as the baseline. The question was: Should the reductions be implemented through taxes on the emissions or through a market for emission permits? In other words, should the inspiration come from Sweden or from the United States? The U.S. delegates made it clear that the only way to have an agreement approved by the U.S. Senate was to adopt the U.S. model—that is, create a market for emission permits across Annex B countries.

The irony of the situation is that these same U.S. delegates knew that no agreement would be approved by the Senate. (According to the U.S. Constitution, no international treaty may be ratified without Senate approval.) Indeed, in July 1997, a couple months before the Kyoto Conference started, senators Robert Byrd (D-WV) and Chuck Hagel (R-NE) sponsored what became known as the Byrd-Hagel Resolution, which stated,

> The United States should not be a signatory to any protocol to the United Nations Framework Convention on Climate Change of 1992, at negotiations at Kyoto in December 1997, or thereafter, which would mandate new commitments to limit or reduce greenhouse gas emissions for the Annex B Parties, unless the protocol or other agreement also mandates new specific scheduled commitments to limit or reduce greenhouse gas emissions for Developing Country Parties within the same compliance period.[5]

The resolution passed 95–0, a rare example of unanimity in Congress. Hence, the Kyoto Protocol, with its mandatory emission reductions for Annex B countries only and its ambitious market architecture, was effectively dead before it was ever made public.

The market would never have properly functioned anyway. The regulating and stabilizing action of the EPA has been a critical factor in the success of the U.S. SO_2 market, but no equivalent public institution is considered in the Kyoto Protocol, as the member nations were all reluctant to have their sovereignty significantly dented. European countries have tried, with various degrees of success, to stick to their commitments as defined in the protocol—for example, by organizing a market for emission permits under the authority of EU institutions. However, despite a level of supranationality in the EU unparalleled elsewhere, the market has been poorly steered, has suffered large-scale fraud and cybertheft, and, above all, has been crushed under a glut of permits. Indeed, there has been a chronic overallocation of permits in an attempt to appease the large emitters involved in the market. Finally, with the unfolding economic crisis from 2008 onward, demand fell to unprecedented low levels, and prices plummeted—to less than €5 in January 2013. (That was not an isolated accident as the performance of the market had continuously worsened for the previous two years.) Thus, it is now as if CO_2 emissions have no effect on climate or as if climate change is irrelevant. The whole story is a disaster—not so much because it resulted in greater European CO_2 emissions (they are actually rather modest in the context of world emissions) than would have been the case with the market functioning as initially envisioned, but because it jeopardizes the perspectives of CO_2 trading at the global level. What could possibly be done with the institutional mess at the global level that could not be done within the relatively strong EU institutions?

In the United States, regional markets for CO_2 permits have tentatively been set. In 2005, the governors of nine Northeast and mid-Atlantic states started a market for CO_2 permits as an instrument to cap CO_2 emissions from power plants. The results obtained during the first three years were encouraging (emissions went down 23 percent in that time), but later, they were less so. Similarly, in 2012, the California Air Resources Board was established to control a broad cap-and-trade scheme, broader in terms of the number of sectors concerned. Its institutional setting is more coherent and stronger than has been seen elsewhere. Quebec has since teamed with California, and Ontario decided to join in 2017.

SWEDEN, AGAIN: LEADING THE PACK OR RACING ALONE?

In Sweden, the tax on CO_2 emissions started at €27 in 1991 and was increased first to €40 in 1996 and then to €100 in 2004. It has been €117 per ton emitted since 2009 (with always 50% rebates for industries exposed to international competition). During those two decades, the total amount of annual taxes (all taxes, not just pollution tax) decreased, from 55 percent to 45 percent of gross national product. All increases in the tax on CO_2 emissions have been announced well in advance to facilitate adjustments. Recovering the tax proceeds is not costly, as the tax is collected upstream from approximately 300 economic agents that introduce the products concerned (mainly fossil fuels) into the national economy.

Norway and the Netherlands are following suit in their tax plans, as are Denmark and Finland—albeit, at some distance. However, viewed from most other countries, the Swedish rates (even the reduced ones) seem ludicrous. Notwithstanding, Sweden and its followers are among the most efficient, open, and prosperous countries in the world, with relatively low unemployment rates and public deficits and high levels of welfare. To say the least, pricing carbon at serious levels doesn't seem to jeopardize competitiveness.

In Sweden, the distributional effects of carbon taxes have been addressed first within the restructuring of the entire fiscal system (Sterner 2011) and then by reorganizing land use and public transport and by promoting untaxed fuels such as wood pellets and waste for heating. The CO_2 tax is an essential, but far from unique, economic instrument used to reach the Environmental Quality Objectives adopted by the Riksdag. These objectives are the expression of a broad political consensus and, as such, are supervised by the All Party Committee on Environmental Objectives; the Swedish Environmental Protection Agency is responsible for their implementation. According to the objectives, by 2020, GHG emissions should be down

40 percent from their 1990 level. By 2011, they were down 16 percent, and most of that decrease had been achieved since 2000. During the same period (1990–2011), gross domestic product was up 58 percent. By 2020, fossil fuels will no longer be used for heating, and by 2030, they will not be used for transport.[6] By 2050, residual GHG emissions should not exceed 10 percent of the 1990 level. Of course, Sweden is not the only country committed in principle to achieving spectacular environmental objectives. It is, however, among the rare ones to have resolutely put itself on a trajectory effectively aiming at such objectives.

It obviously makes sense for Sweden to tax SO_2 and NO_x, as well as other factors of those pollutions directly affecting them. Doing so significantly improves the country's environment, the balance of its public finances, and its economic efficiency and welfare. But what about taxing CO_2 emissions? These taxes also contribute to the balance of public finances and to economic efficiency insofar as they induce cost-reducing substitutions and innovations, some of them liable to open new markets. But they might also create discomfort in personal lives and difficulties for firms (however reduced, the rates they pay are far from negligible). Why should Sweden impose upon its citizens in this way? If you keep in mind that there are 34 Americans (U.S.) and 148 Chinese for every Swede, you recognize that whatever Sweden does to save the climate cannot be of much direct help to the world. The spirit of the Swedish endeavor is encapsulated in the following resolution by the Riksdag:

> The overall goal of Swedish environmental policy is to hand over to the next generation a society in which the major environmental problems in Sweden have been solved, without increasing environmental and health problems outside Sweden's borders.[7]

As far as climate is concerned, however, the future of Swedish children—and of all children in the world—hangs very much on what China and the United States do or do not do and very little on the effort made by Sweden. Yet, neither the United States nor China seems to have objectives matching those of Sweden. As President George H. W. Bush stated at the Rio Earth Summit in 1992: "The American way of life is not negotiable." That declaration still reflects the feelings of many Americans.[8] Likewise, Yu Qinqtai, China's special representative for climate change negotiations from 2007 to 2010, said in a talk at Peking University's School of International Studies, "We cannot blindly accept that protecting the climate is humanity's common interest—national interests should come first." In China, however, this perception of national interest seems to be coming more in line with "humanity's common interest."[9]

The Swedes do not look completely rational—in the narrow sense economists are fond of—but they are keen to make their experience available for inspiration when China, the United States, and others panic into fundamentally changing their ways—with the risk that it might be too late.

FINANCING: FROM PERVERSE SUBSIDIES AND INVESTMENTS TO NEW CONCERNS ON FINANCIAL MARKETS

FINANCING THE DESTRUCTION OF NATURAL CAPITAL

According to the International Energy Agency's (IEA's) *World Energy Outlook 2015*, in 2014, subsidies to fossil fuel production and consumption amounted to about $500 billion; however, those subsidies are on a decreasing trend, linked to falling prices, especially of oil. The implicit subsidies, which correspond to the external costs inflicted by both production and consumption of fossil fuels, add to these direct explicit subsidies; therefore, the IMF estimates are greater by one order of magnitude (Coady et al. 2015). In that same year, direct subsidies to renewable energy sources amounted to about $120 billion, which could easily be dispensed with if fossil fuels were priced at their true costs.

Subsidies to fossil fuels undermine efforts toward improved energy efficiency; they are detrimental to the environment and the climate; and they are a drain on public finances, particularly in developing countries. Consumption subsidies benefit disproportionately middle-class urban dwellers and only marginally benefit poor households

(Sterner 2011). What poor households get in return is not insignificant for them and cannot be removed without compensation; in Indonesia, where the share of the state budget spent on fossil fuel consumption subsidies has significantly decreased since 2008, unconditional cash transfers have been targeted on the worst-off 30 percent of the population.

Although most production and consumption subsidies are wasteful and should be removed, connection subsidies have proved valuable in many places, in Africa, Asia, and Latin America. These subsidies help disadvantaged communities access more convenient forms of energy. For example, the Chilean experience at promoting rural electrification in this way is often mentioned as a positive example.

Developed countries have their fair share of wasteful subsidies. One of the most perverse examples is found in the United States and, to a lesser extent, Europe, where first-generation biofuels, such as ethanol, are heavily subsidized, even though they require for their production more fossil fuel than they save when used in vehicles. These biofuels also compete with food production (see chapter 3). More generally, subsidies amounting to hundreds of billions of dollars, coupled with privileged access to credit, contribute to the destruction of natural resources and ecosystems—in particular, in the mining industries, the fishing industries (see chapter 1), and agriculture (see chapter 6). Each year, more than $700 billion of public money goes to activities that destroy natural capital; moreover, these activities escape paying for the even larger—by a wide margin—"external" costs they inflict.

It is not merely a matter of subsidies. Some banks have recently decided to discontinue financing projects centered on coal, from mining to burning it in power plants. However, there is still no shortage of loans—often backed by government agencies in China, the European Union, and the United States—nor of direct investments in the production and large-scale uses of fossil fuels—coal and oil in particular. Banks also massively support the downright destruction of key ecosystems (for a striking example, see box 9.1).

According to the IEA's *World Energy Outlook 2016 Executive Summary*, "During the period 2000–2015, close to 70 percent of total

BOX 9.1 BIG BANKS SUPPORT FIRMS LINKED TO DEFORESTATION

The link between the world of finance and the destruction of forests became clearer with a recent study that found that global banks, including three large banks in Indonesia and international banks such as China Development Bank, HSBC, JPMorgan Chase, Mizuho Financial, have been financially supporting forest-risk companies. The study "Forest and Finance" (Rainforest Action Network, TuK Indonesia, and Profundo 2016) revealed that, between 2010 and 2015, financial institutions provided more than $38 billion worth of commercial loans and underwriting facilities to fifty companies implicated in deforestation in the Asia-Pacific region. Banks provided loans and underwriting facilities to companies through their production and primary processing operations in four sectors: palm oil, pulp and paper, rubber, and tropical timber.

The study was drawn up by the community group TuK Indonesia, California-based Rainforest Action Network, and Dutch consultancy Profundo.

supply investment went to fossil fuels" (2). The agency anticipates an improvement for the period up to 2040: "A cumulative $44 trillion in investment is needed in global energy supply in our main scenario, 60 percent of which goes to oil, gas and coal extraction and supply, including power plants using these fuels, and nearly 20 percent to renewable energies." Needless to say, these prospects are incompatible with the goals set in the Paris Climate Agreement; if these goals were followed, climate change would indeed take a catastrophic turn.

To somewhat correct this huge imbalance, Espagne, Aglietta, and Fabert (2015) proposed that central banks should develop specific procedures toward redeeming loans that commercial banks made to finance projects contributing to an overall reduction in GHG emissions. Under this proposal, a public body would be put in charge of verifying and certifying the reductions obtained. It would then issue certificates representing a fraction of the value being transferred to the central bank. In due course, the bank might obtain reimbursement of the certificates from the state budget. The justification for this chain of operations is that the value of a certificate is the value of the contribution of the corresponding investment to the reduction of a "public bad" (the opposite of a public good)—in this case, the stock of GHGs in the atmosphere. This mechanism would also function as an incentive for the state to introduce, and progressively increase, prices on GHG emissions—on CO_2 emissions, in particular. This incentive would have the effect of shrinking the base of the certificates. Indeed, the more you make polluters pay for their emissions, the less you have to pay for getting the level of reductions you are after.

For the time being, however, there is no such involvement from central banks. Although so-called green bonds are issued, there is no systematic verification process of their destination. In China, however, there are moves toward a system that is akin to the one mentioned above, involving the People's Bank of China. Similarly, the EU is working out a certification procedure, though not in cooperation with the European Central Bank.

THE DIVEST-INVEST MOVEMENT

More substantial advances have been seen on financial markets, where investors face a climate risk that derives from both the effects of climate change and the effects of public policies, regulation rules, judicial decisions, and citizen initiatives meant to contain and reduce climate change. Markets tend to react to what is perceived as threatening certain classes of assets or as boosting others. Warnings against so-called stranded assets (see chapter 3)—in particular, those warnings by Mark Carney, governor of the Bank of England, in his capacity as chair of the Basel-based Financial Stability Board—have a particular resonance.[10]

Until recently, most investors had not properly factored in the climate risk, with a large number simply ignoring it. When a risk is broadly underestimated on the markets, it is not adequately rewarded. Those investors who happen to take it seriously try to rebalance their portfolios away from high-carbon footprint assets (like coal, oil, or cement producers) and toward assets positively associated with the transition to a decarbonized economy (like renewable energy providers or

BOX 9.2 DIVESTMENT-INVESTMENT DYNAMICS

Between September 2014 and September 2016, the potential of portfolio decarbonization has been multiplied by a factor of more than 40, and institutional and private investors with $3.4 trillion of assets in management have joined the divest-invest movement. Among them are insurance companies (e.g., Allianz Life and AXA), pension funds (e.g., New York State Common Retirement Fund, California Public Employees' Retirement System [CalPERS], California State Teachers' Retirement System [CalSTRS], Second Swedish National Pension Fund AP2), sovereign funds (e.g., Norway's sovereign wealth fund, France's Caisse des Dépôts et Consignations), and foundations and endowments (e.g., Divest-Invest Philanthropy Coalition, University of California, University of Oxford). The Norway fund and the Californian retirement systems have been ordered to move by their respective legislatures.

producers of energy-efficient materials). Pioneers like the Stanford University Endowment or the Rockefeller Brothers Fund have kick-started this "divest-invest" movement, which has quickly gained momentum (see box 9.2).

Most investors, especially portfolio managers, are determined to avoid shifts that would significantly reduce financial returns during the period before the markets properly factor in the climate risk. These investors tend to adopt strategies consisting of hedging the climate risk, subject to an upper bound on any reduction of financial returns. In "Hedging Climate Risks," Andersson, Bolton, and Samama (2014) described how to maximize the decarbonization of a portfolio—that is, to minimize its carbon footprint—under the constraint that the loss in returns must be less than an acceptable upper bound (typically less than 1 percent). They suggested progressively offloading assets with larger carbon footprints and simultaneously reoptimizing the portfolio in terms of returns.

Standard benchmarks, such as the S&P 500 in the United States or MSCI in Europe, which are tracked by numerous investors, have generated variants along these lines. S&P has generated the S&P/

IFCI Carbon Efficient Index, and MSCI has generated MSCI Europe Low Carbon Leaders Index. These new indices provide 60–80 percent decarbonization rates, while keeping track of errors (return losses) under 0.8 percent. When climate risk is properly factored into the markets, the new indices will start to outperform the corresponding benchmarks, as is already happening as coal assets stall.

WHERE PRIVATE FINANCE AND PUBLIC POLICY IN TANDEM PUSH FOR THE TRANSITION

When investors have rebalanced a portfolio, they will benefit from any circumstance that increases the degree to which the climate risk is factored in on the markets. They will thus benefit from public policies that either increase the awareness of the risk or directly penalize carbon content or effects, such as carbon pricing under its various forms. Investors will then become part of a broadening constituency in favor of more ambitious policies for containing and reducing climate change. Investors get to push for more ambitious policies, and more ambitious policies legitimate their portfolio choices. When that happens on a significant scale, public and private finance might join in a climate-friendly dynamic.

10

GLOBAL GOVERNANCE OF SUSTAINABLE DEVELOPMENT

THE GREAT ENVIRONMENTAL TRANSFORMATION AND INTERNATIONAL RELATIONS

Sustainable development, as a combination of environmental protection, economic growth and social equity, has become a planetary issue. Although it is still somewhat disjointed and multiform, it has undoubtedly become an imperative, at least in the rhetoric of international, national, and local actors and institutions. This imperative, this collective norm, has a history that is simultaneously very recent and already a bit old. It is, on the one hand, deeply rooted in a critical discussion of modes of production, consumption, and distribution of goods and resources that has led to radical contestation, a fundamental change of paradigm, and a new development model. On the other hand, it has evolved along scattered and scalable coordination patterns that emphasize careful experimentation and learning rather than prescription.

Sustainable development governance has a double nature. It has a substantial nature through the radical criticism it voices and embodies. But it also has a procedural nature through the learning it implies for numerous political, economic, and scientific actors and institutions. This procedural nature has transformed it over time. Indeed, sustainable development means not only a critical reflection but also a search for solutions—building coordination for collective action, creating institutions, establishing global mandates, and evaluating results and procedures.

The environmental issue emerged in the international relations field in the late 1960s and early 1970s. The 1972 Stockholm conference is commonly considered the first internationalization of questions that were previously thought of as essentially regional, national, or local issues. During this conference, some problems—such as the negative impacts of capital-intensive agriculture, the extinction of animal species, transboundary pollution within a watershed or a highly industrialized area, and natural resource degradation (resources that were still thought to be abundant and available on a global scale)—began to shift from a localized, sector-specific status to a global one.

In Stockholm, some front-runners, or what can be considered the "environmental avant-garde," succeeded in mainstreaming a fundamental principle that is still the basis and justification of international action and coordination today—that is, the earth's resources are not inexhaustible. All the disorders, poor management systems of resources, and polluting activities are the elements of a single international collective action problem: the uncontrolled use of natural resources by the human species. This recognition of the problem's collective nature laid the foundations for what was not yet referred to as "global governance." The title of the seminal report prepared for the Stockholm conference, by Barbara Ward and René Dubos, expressed with clarity this new concept: "Only One Earth: The Care and Maintenance of a Small Planet," implying that, since we have only one planet to live on, we had better work together to maintain our means of existence (Ward Jackson and Dubos 1983).

However, it took exactly twenty years—a time marked by the prominence of macroeconomic questions, including inflation, sovereign debt, and foreign exchange—before the Stockholm appeal found the attention it deserved. The Earth Summit held in Rio de Janeiro in 1992, where 178 delegations and 110 heads of state convened, finally carried on the perspective initiated two decades earlier. The summit "institutionalized" the environment and sustainable development at the international level. With its three conventions (on climate, biodiversity, and desertification), it presented a first draft of a road map. Since Rio, the production of coordination frameworks started to speed up: some major multilateral environmental agreements were signed (between 500 and 700, depending on definitions), and the number of conferences held to settle their implementation, as well as the number of monitoring reports, boomed. Extensive research measured and modeled resources and environmental evolutions in order to support political decision making and formulate recommendations.

Between the fall of the Berlin Wall in 1989 and the attacks of September 11, 2001, the world experienced what seemed to be a "golden era"—or, rather, a "golden interlude"—of international relations, characterized by a certain confidence in science and rationality. Above all, there was a confidence in negotiated norms and laws to effectively guide coordinated action—in other words, there was confidence in multilateralism. During this "golden" period, the percolation of environmental issues was straightforward and wide-ranging: the impact was visible on the topics of coordination, the actors involved, and all levels of intervention.

After being established as a global issue in Stockholm, the environment gained its true crosscutting dimension during the Rio summit. The summit legitimated government action beyond control of pollution, recognizing that environmental policies were intrinsically sustainable development policies: they affect or even transform economic models, modify income distribution, and therefore sort out new winners and losers. This general feature of environmental policies is, as the first twenty years after Rio demonstrated, the core difficulty involved in their implementation.

More than fifty years have passed since the first timid recognition of the environmental issue as a collective action problem requiring international coordination. It has taken all this time to conclude a truly global agreement on GHG emissions reduction agreement; this was reached at COP21 in Paris and is to be implemented in 2020. The fact that it took so long shows the difficulty of environmental governance to accomplish the transformations called for during the Stockholm and Rio summits, despite the tremendous proliferation of institutions and initiatives. As the Danish minister of foreign affairs, Carsten Staur, highlighted in the preface to a study published by the International Institute for Sustainable Development (IISD): "It is difficult to imagine how to overcome the paradox of institutional success and environmental degradation" (Najam, Papa, and Taiyab 2006, iv). As the authors of the study indicated:

> Given the reality of increasing carbon emissions, dwindling forest cover, declining fish stocks and disappearing biodiversity, it is clear that while the system of global environmental governance has grown in size and scope, it has not been entirely effective in achieving its larger goals of actually improving the global environment, of achieving sustainable development or even of reversing the major trends of degradation. (14)

They also emphasized that, "in fairness, it may be too soon to seek such results from a system that is still evolving." (idem)

The apparent paradox that characterizes global environmental governance relies on the double nature of its contemporary construction:

- It is a fundamental, crosscutting, multiform issue. The capacities it requires are not necessarily in the core competencies, prerogatives, or mandates of the institutions and actors whose coordination is absolutely required. When appropriate, it also entails new rules, new coordination modalities, new institutional and regulatory constructions, and new deliberation frameworks. While favoring interplay between preferences to a power game, it holds the promise—or, for some, the threat—to deeply modify the classical models of international relations, based on a balance of power and interests between sovereign states, which were (and still are, in many cases) the fundamental pillars of the international system.
- Simultaneously, the environmental question has also found its place in existing institutions and frameworks. Thus, while striving to slowly change them, it cannot avoid falling into their routines and sluggishness (even though it has also made use of a certain efficiency gained in dealing with issues such as security, trade, and health). The states, while officially recognizing the

consequences of the issue's novelty on their relations, as well as a transformation of their sovereignty toward a more shared sovereignty, have abandoned neither their power prerogatives nor the defense of private, national, and multinational interests, which they still support and legitimate.

These two constituent features of global environmental governance eventually led many to consider it as unstable and inefficient, or at least as underserving the challenges and issues at stake. However, one could also argue—notably, when considering its youth and the radical reforms of international cooperation it seems to require—that environmental governance is still a remarkable and unparalleled learning process, which must navigate between the substantial issues and their urgency, on one hand, and the decision-making processes and implementation of solutions, on the other.

It is true that the environmental question is still dominated or sidelined by (not to mention under the constant criticism of) realists, for whom it is bound to remain a secondary and subordinated issue, with much less weight than interests or power. However, the mere existence of the differentiated ways of coordination that have been established in more and more complex and detailed regimes challenges theories that are focused only on power relations or on the inertia of path dependencies.

A RADICAL CHALLENGE OF THE ESTABLISHED MODEL

The recognition—and then the rising importance and pressure—of environmental issues in the international agenda has imposed itself within a double dynamic: on one side, a social movement and a scientific and economic criticism of the international order born of the end of World War II, and on the other, the perception of negative impacts of economic positivism (what has finally been termed *productivism*). The features of this double dynamic, and above all their convergence, explain a great part of this governance system, as well as the fact that it finally has become a question of general interest even now as a question of high-end politics after being considered restricted to narrow circles of specialists.

SOCIAL AND SOCIETAL PROTESTS . . .

The environmental question invited itself into the great social movements of the 1960s and 1970s with the rise in power of political ecology, which was echoed by the creation of several international environmental NGOs, such as the World Wide Fund for Nature (WWF; now the World Wildlife Fund) in 1961, Friends of the Earth in 1969, and Greenpeace in 1971. Some major environmental disasters, such as the first major oil spill in 1967 (the *Torrey Canyon* sinking near the Breton coast) and the chemical pollution of the Rhine over more than 600 km in 1969, led to the death of several million fish (not to mention many more localized disasters) and contributed to an increase in public and political awareness of the necessity to face environmental issues.

The emergence of the environmental debate did not rely only on these disasters. Student and youth movements, which were a central part of the 1967–1968 protests, were also a societal movement in which the debate over the consumer society took an important place, before becoming politicized into a right-versus-left or conservatism-against-revolution divide. The movement's ideological base was sensible to environmentalists' arguments aiming to emancipate themselves from the confined naturalist and nature conservationist circles and to connect the defense of the environment with societal causes, notably in developed and democratic countries but also in several developing countries.

The concept of the environment, which was starting to attract wider attention, clearly suggests this shift from a nature-protection stance to a broader understanding of relations between human societies and nature. Then the social movement fighting for justice, human rights, and peace, together with anti-consumerism values could integrate the recognition of the importance of the environment. This recognition was supported by numerous scientific disciplines that have provided—year after year—more and more converging and strong arguments substantiating the priority of the environmental issue, as well as its central place in collective concerns. Science and social movements began to connect in a process that has seen many scientists joining activist movements when confronted by the inertia of the political decision making.

. . . AS WELL AS SCIENTIFIC WARNINGS

In 1962, U.S. biologist and conservationist Rachel Carson wrote *Silent Spring*, in which she described the impact of intensive agriculture and the massive use of petrochemicals on the environment—in particular, on the food chain. The title calls to mind a spring without any birds singing after their extinction due to pesticides. Carson's book attracted tremendous attention after its release in the United States and gave rise to an environmental movement independent of existing conservationist large groups (such as the Sierra Club, founded in 1892) and the creation of national parks (such as Yellowstone in 1876 and Yosemite in 1890). In fact, soon after the publication of *Silent Spring*, the President's Science Advisory Committee (1963) submitted a report on the use of pesticides. Although the report highlighted the benefits of chemical products for agriculture, the members of the expert panel noted that their use could prove dangerous to human health and should be better controlled. This concise report laid the foundations of the U.S. environmental legislation, most notably through the creation of new agencies; it also led to a definitive ban on the use of DDT in 1972.

At the time, there were increasingly discordant voices heard in the field of economics. U.S. economist Kenneth Boulding (1966) stressed the uniqueness of the earth and its limits and urged a more rational use of its finite resources. In 1973, another U.S. economist, Herman Daly, published *Toward a Steady-State Economy*, which revived the concept of the steady-state economy introduced by Robert Thomas Malthus (1836), giving it a positive meaning: growth should not be the ultimate goal of an economic system. Meanwhile, biophysicist Nicholas Georgescu-Roegen (1971) warned of the risk of overexploited resources unable to regenerate themselves, building on the law of entropy. His work has been at the origin of the debates on the limits of unbridled material economic growth, and thus on the concept of "degrowth," which he was the first to mention and advocate for.

More formally, in 1970, the Club of Rome, a think tank of scientists, economists, and national and international officials, commissioned a report written by four researchers from MIT: Dennis and Donella Meadows, Jørgen Randers, and William Behrens (1972). It was the first major study highlighting the ecological threats of economic and demographic growth; it even considered that economic growth might one day have an end. It also marked the first effort to simulate the interactions between human activities and their environment, which was made possible thanks to advances in computer technology. Because of the rather simplistic assumptions it was based on, the "Meadows report" caused great controversy when it came out. It was sharply attacked and vigorously contested by the productivists, regardless of their origin: from countries with market economies to the socialist bloc and the Third World. Nevertheless, its impact has been fundamental in that it was the first time that economic science essentially said: "Beware, this system cannot be perpetuated indefinitely because it consumes too many nonrenewable resources way too fast and because it releases too many polluting substances into the environment."

Since these early movers, research from the field of earth and life sciences was developed to document the interactions among human activities, ecosystems, and geophysical balances. As we saw in chapter 1, there has been convergence of diagnoses from several fields, all stating that human activities carry systemic risks and have already led to irreversible changes. Paul Crutzen, winner of the Nobel Prize for chemistry, even argued that we have entered a new geological era, the Anthropocene, in which human activities have become the main factor behind the evolutions of physical and biological environments.

The Meadows analysis has been revised since it came out—notably, by Johan Rockström and colleagues (2009), who floated the idea of planetary or biophysical boundaries, which, if exceeded, could have disastrous effects on humanity. This is also the idea behind Oxfam's concept of the "doughnut"—a visual framework that added to the superior physical boundaries the complementary concept of social boundaries, creating a space between the two in which one human must remain (Raworth 2012).

As in 1972, today science continues to propose measures to address problems and concepts to frame political debates. This can be observed, for instance, in the title of the Union Environment Action Programme to 2020 of the European Commission—"Living Well, Within the Limits of Our Planet"—which echoes Rockström's paper just as much as it does the Meadows report. The similarity of the words employed today and forty years ago indicates that the problems have essentially remained the same, while governance keeps going in circles: the environmental challenges still call for the same urgent need for action. However, even though some positive cycles in international governance can be observed—as in the case of the 1990s "golden interlude"—we are currently not in a situation comparable with the one in which the Meadows report was written. The acuteness of our knowledge of environmental problems has extended and deepened, thanks not only to developments in science, but also to the fact that the number and magnitude of these problems has increased. Trade liberalization helped developing—and particularly emerging—countries catch up with developed ones. This "catch-up" continued to accelerate during the turn of the century, most notably with China's accession to the WTO.

This globalization hastened convergence across countries of gross domestic product, based on intensive use of natural resources and to the detriment of protection of the planet. Indeed, while awareness was rising, the actual trends of destruction of natural resources accelerated: more importantly destruction of natural resources was and still is considered by many as the unintended but necessary condition for growth.

Slowly the idea of the compatibility between economic growth and planet protection is emerging, but the idea still needs more empirical evidence to drive change, such as the reduction in CFC and SO_2 emissions since the turn of the century (Antweiler, Copeland, and Taylor 2001) or the decoupling of economic growth and CO_2 emissions in a number of countries since 2009.

The global governance of sustainable development is at the center of multiple pressures. Governing the transformation of economic development toward sustainability, if taken seriously, is a radical project implemented by institutions originally unacquainted with it.

A NECESSARY COLLECTIVE ACTION

Granting that the environmental question (in all its diverse elements, including climate engineering, biodiversity, natural resources, and pollution) is a planetary one and an international collective action issue, the central question is how to agree on a conceivable action to identify, evaluate, and deal with the threats to the environment and, in the long term, how to imagine and test a production, transformation, exchange, and consumption model that is environmentally sustainable, socially equitable, and economically viable.

The structure of the system in which this question emerges—that is, our contemporary world, made more interdependent after four waves of globalization—inevitably raises the question of international coordination, independent of the substantial and inherent problems of collective action for the environment. The current context of globalization is a specific context for collective action, with its rules, principles, and law—and perhaps its transformation of power relations, induced by the trade of goods, services, and capital over time. The global environmental governance cannot be designed without taking this context into consideration. Environmental collective action has come to the top of an already existing collective action structure, requiring reforms of the previous system in order to be more efficient. The radical project of sustainable development actually raises a "meta collective action" problem, restricted by the resistance of institutions.

Another source of resistance comes from the states themselves. We need the intergovernmental scale in order to manage global problems, but these problems actually transcend these same governments and are only partially in their established and well-understood field of competencies. This fundamental contradiction has distorted intergovernmental negotiations for forty years: the combination of national interests and powers, from a political economic sphere well-known by the governments, limits and hampers action in favor of common goods.

The repeated inability to create a World Environment Organization (WEO) since the 1992 Earth Summit in Rio, where it was mentioned seriously for the first time, can be seen as evident resistance from states to limit their negotiating edge and sovereignty. Such an organization would, according to its critics, only be a useless level of meta-coordination. Between the lines of such criticism, however, we can find the implicit fear of a universal organization, able to gather the works of many different competent organisms and regimes and, more important, to mark out and reduce the negotiating edges of the most powerful states' economic and strategic interests—all in the name of a collective approach transcending the specific interests of nations.

Except for inducing a "pareto-improving" shift of the preexisting balance of cooperation and satisfying all the parts without harming anyone, environmental cooperation between states is all but undemanding, yet it is necessary. Why would states want to cooperate? In what name, on whose behalf, and based on what kind of legitimacy would they do so? In the light of traditional international relations theory, states only cooperate when the balance or unbalance of powers composing the international system constrains them to do so. Hence, such cooperation would require a dominant (hegemonic) state to rally a number of critical countries to its cause and to impose the view that the environmental question is an issue of national security and interest. It would then have to enforce this idea as a position of common good at the international level. As an example, in 1948, the United States and the United Kingdom succeeded in imposing the idea and later the creation of GATT after the UN Conference on Trade and Employment in Havana. If we disregard the exceptional case of chlorofluorocarbons (see box 10.1), we are certain to wonder which power or club of powers could be capable of doing the same for the sake of environmental protection today.

It is neither desirable to define some kind of dictatorship (in the Roman sense of the term), even one based on rational arguments that would impose "one vision of the common good," nor plausible to think of a global authority that is not subject to the perpetual dissension and competition between states. The collective action necessary to handle environmental issues seems to remain a puppet in their hands: it is permitted and created by the states and, at the same time, weakened and limited in its actions by their goodwill. But far from exhausting states' sovereignty, global environmental governance

BOX 10.1 HEGEMONY AND COOPERATION: THE CASE OF CHLOROFLUOROCARBONS

The example of the Montreal Protocol shows that the model of action led by a dominant actor is possible and can be efficient, even when in a peculiar setting. The Montreal Protocol, signed in 1987, aimed to phase out the production of some substances responsible for ozone depletion, such as chlorofluorocarbons (CFCs). The United States played a key role in proposing the protocol and getting it adopted by all countries. The U.S. government had support from its domestic chemical industry, which was dreading a national prohibition that would have harmed its competitiveness with international competitors not affected by the same regulation. The idea of an international treaty progressively phasing out the production of CFCs meant that the same standards would be applied to all. To some extent, it even gave an advantage to U.S. producers, who had already found some substitutes for CFCs (Morrisette 1989).

The protocol represents the prototype of a limited agreement, rapid and efficient in the history of environmental international negotiations, whose great strength is its flexibility, which allows swift integration of new scientific information. All the stakeholders of this agreement consented to create panels on science, environment, and economy and committed to regularly reviewing the different provisions of the agreement, taking into account the successive advances of the different panels.

represents an opportunity to transform and reaffirm the latter and even extend it into new areas, if necessary, by a delegation of power through a legal agreement of little binding nature.

However, states cannot easily use their authority to limit the involvement of scientific communities, political representatives, economic sectors, and societies in environmental issues. The example of China is a vivid illustration of this fact. Since citizens in big cities in China have real-time access to information about air quality, the Chinese government is considering progressively eliminating the use of coal and, by so doing, reducing the carbon emissions induced by electricity generation. No one could have foreseen this political evolution only ten years ago.

If we exclude the preexisting cooperation framework—especially on economic matters—the critical issue of collective environmental action can represent a strong rationale for international cooperation to preserve resources facing threats of deterioration, exhaustion, or extinction. This rationale calls for the overcoming of the tragedy of the commons as described by Garrett Hardin (1968):

> Therein is the tragedy. Each man is locked into a system that compels him to increase his herd without limit—in a world that is limited. Ruin is the destination toward which all men rush, each pursuing his own best interest in a society that believes in the freedom of the commons. Freedom in a commons brings ruin to all. (1244)

The defects and failures in the management of natural resources can be explained by the quest for personal immediate gain, which compromises the long-term use of a limited common good (the mad rush after the last fish in the management of fishery, for instance). Hardin suggests two types of solutions to overcome this tragedy: either the privatization of the resource as an incentive for a more rational management by owners, or the nationalization by the state to regulate access to it in the name of the common good.

Elinor Ostrom (1990), who based her work on numerous empirical studies, later described a third way: the management of common goods (common-pool resources) by local communities through social norms and institutional arrangements. However, even if scholars from the field of collective action theory have documented the collective management of common resources, it mostly applies to localized common goods, for which the creation of appropriated institutions is possible due to the limited number of stakeholders. Ostrom's assumption that it is possible to transfer common-pool resources on the local level to global environmental public goods has not found any satisfying evidence for two main reasons. First, global common goods have, by definition, no other limit but the earth and all of its population and it is not possible to define precisely responsibility and ownership of these goods, nor to exclude easily anyone from their consumption, use, or benefit: examples are many, like the fish stocks in international waters, or CO_2 sequestration by forests as a climate regulator. This lack of excludability makes it difficult to implement the co-responsibility or common management supported by Ostrom, which needs to exclude free-riding behavior. Second, global common environment goods—at least the institutional arrangements, which are supposed to design and implement the rules of collective action to provide them—are still in the making; thus, there is no defined set of stakeholders "sharing" the responsibility of a common pool of resources.

That being said, the "Hardin-Ostrom" assumption, though out of sync with the power-relation patterns and social representations of environmental questions, continues to fuel the process of coordination at the international level. The construction of this international collective action is—implicitly or explicitly—based on the possibility of limiting the "prisoner's dilemma," a situation in which the confrontation of different individual interests does not necessarily lead to a collective optimum through the negotiated elaboration of sanction-and-reward systems. The prisoner's dilemma also highlights the importance of issues of trust and loyalty between actors in and for collective action. This is why regular negotiation rounds (e.g., conferences between stakeholders within international negotiations on climate or biodiversity) and the constitution of different working groups and committees of experts contribute to higher levels of trust among states and have the potential to enhance the predictability of behaviors during negotiations, if not to reach the "pareto-improving" result that should be induced by resolution of the prisoner's dilemma. These repetitive processes often seem to be useless because of their slowness and their not particularly spectacular results. A more constructivist assessment of the rationality behind collective action, however, far from making it a costly means for achieving uncertain results, gives it a primary place in the construction of a common vision and framework. These long processes create trust and modify, through the construction of common knowledge, the perception of interests. There is an interplay between preferences and the power game: governance seems to remain this two-faced Janus that always seems to grimace, no matter how you look at it.

DIFFERENTIATED COORDINATION MODES

In an attempt to organize collective action around "global commons," we may simplify and reduce the different modes of coordination to three major functioning modalities: international organizations, multilateral agreements, and the coordination of good practices or the creation of networks among actors and initiatives (mostly non-state ones). These three modalities have emerged during different periods. Their succession (and coexistence today) relates to the evolution of ideas; but it also reflects a process driven by a quest for better results: the apparent inefficiency of one modality seems to have induced the deployment of a new one. This trial-and-error process is a typical feature of environmental governance.

THE ATTEMPT TO CREATE AN INTERNATIONAL ORGANIZATION

On June 16, 1972, the UN Conference on the Human Environment in Stockholm stressed the "need for a common outlook and for common principles to inspire and guide the peoples of the world in the preservation and enhancement of the human environment" (1). After four years of preparation, the conference reaffirmed the UN model of large specialized international organizations with the creation of the UN Environment Programme (UNEP).

At the end of the 1970s, as a result of the relative failure of UNEP in imposing the environmental question on the international scene, new forms of arrangements between states emerged. The goal was not to create or mobilize a dedicated international institution, to which a delegation of sovereignty is granted for the definition and resolution of a given environmental problem, but rather to focus on ad hoc negotiations of agreements between sovereign states. This tendency was fueled by the progressive erosion of the U.S. domination of world affairs, which is slowly giving way to a new system of multilateral balances (see chapter 11) and to an increasing influence of national political agendas with the creation of environment ministries. Between 1971 and 1975, thirty-one major environmental laws were voted on in Organisation for Economic Co-operation and Development (OECD) countries (which is as many as occurred between 1956 and 1970). This movement might seem to be a setback for international cooperation per se, but the creation of national actors and initiatives also witnesses the dissemination of environmental topics and issues into national policies. Indeed, it would be unthinkable today to define a national or regional government without a public ministry dedicated to the environment.

THE PREFERENCE FOR AGREEMENTS BETWEEN SOVEREIGN STATES

In 1990, Robert Keohane (1990, 731) defined multilateralism as "the practice of coordinating national policies in groups of three or more states." This definition was later amended by James Caporaso (1992), who added some more precise principles: "As an organizing principle, the institution of multilateralism is distinguished from other forms by three properties: indivisibility, generalized principles of conduct, and diffuse reciprocity" (1).

The idea of developing multilateral environmental agreements culminated during the Rio Earth Summit in 1992, with the creation of its three major conventions. In the years following the summit, the idea of international governance developed, leading to a reciprocal and negotiated renunciation of a certain form of unilateral sovereignty in order to handle global causes. In this sense, states should then submit to the demands of a superior authority for the supply and conservation of global public goods.

The academic literature provides some precisions on the performance conditions of such agreements. The environmental conventions that should work best are those with limited impact on each country's economic development and should remain confined to technical targets, such as the Convention on International Trade in Endangered Species of Wild Fauna and Flora (CITES) or the Bonn Convention on the Conservation of Migratory Species of Wild Animals. In other words, because of the weakness of the economic and social issues associated with them, the best-performing global environmental agreements are not, strictly speaking, sustainable development agreements. According to Scott Barrett (1994, 2), agreements must be "deep but narrow or broad but shallow," illustrating that there is a necessary trade-off, or even a conflict, between the form and substance of international agreements—or, more precisely, between their scope and depth (Raustiala 2005).

As mentioned earlier, macroeconomic concerns postponed the repercussions of the Stockholm conference. After September 11, 2001, the traditional issues of foreign policy (e.g., economic interests, security) marginalized environmental agreements between sovereign states in the international agenda and put an end to ten years of particularly active diplomatic activity for the environment. The attacks of September 11 brought profound changes to the U.S. perception of international relations: after having strong political leadership in multilateral processes (see Bill Clinton's "Democratic Enlargement" doctrine), the United States put security at the forefront of collective goods.

The second Bush administration developed the idea of an à la carte multilateralism, with case-by-case, rather than general, commitments. In so doing, they gathered "coalitions of the willing" for specific objectives.

> The Bush administration's circumvention of international cooperative norms and institutions has notably included the refusal to cooperate with multilateral efforts to govern the global commons in order to curb the degradation of our global resources and work toward a sustainable future. In March 2001, President George W. Bush declared the Kyoto Protocol "dead" signifying his decision to abandon the treaty completely (the United States is the only major carbon dioxide emitter to have rejected the Kyoto process). (Powell 2003)

A traditional view of international relations, based on the balance of power, then imposed itself, giving priority to a narrower vision of national interests and marginalizing the global governance project (notably, as theorized by the UN Development Programme [UNDP] in 1999), which was rooted in the idea of global public goods (Kaul, Grunberg, and Sten 1999). This is what we call the "realist backlash," which is actually not entirely attributable to the Bush administration (Lerin and Tubiana 2005). Negotiators were already reminded of the principle of reality a few months before the 1997 conference in Kyoto, when the U.S. Senate passed a resolution expressing that it would never ratify any treaty requiring developed countries to make GHG emission reductions, while developing countries were excluded from such reductions.

With this change of U.S. posture, international coordination returned, to some extent, to its historical major concern—security—and its standard operating mode—the setting of objectives and agenda by the hegemonic power.

> Thus, the new American administration was very active in reshaping the agenda of the World Summit on Sustainable Development held in Johannesburg in 2002. Initially planned for the assessment of the progress made during ten years of intensive efforts in the field of environmental agreements, and for the setting of a new ambitious international coordination phase, the Summit was finally confined to the exposition of some voluntary public and private initiatives in favor of sustainable development. Some less constraining targets than those of Rio were finally set, centered on poverty reduction and the protection of biodiversity. (Lerin and Tubiana 2005, 81)

THE FALTERING OF GOVERNMENTAL AGREEMENTS AND THE RISE OF NONSTATE ACTORS

The World Summit on Sustainable Development in Johannesburg signified "the end of a decade that propelled the environment into an issue of global significance, inextricably linked it to development and energized non-governmental organizations around the world to play a more active, participatory role in environmental governance" (Jasanoff and Martello 2004, 10). It followed the 2001 WTO Ministerial Conference of Doha and merely adopted its conclusions, which stated that the liberalization of international trade was a means to achieve sustainable development. The summit also recognized partnerships among the private sector, states, NGOs, and UN bodies as key implementation tools for sustainable development targets. According to von Frantzius (2002): "These partnerships—voluntary, nonbinding partnerships between civil society, business and government actors, largely operating on a small scale on projects related to sustainability—represent a very different sort of global governance, far more decentralized and less governmental, compared with international treaty regimes, the dominant mode of global environmental governance to date" (17). The Johannesburg targets thus endorsed the idea that interstate agreements were inefficient and that they should rather promote voluntary approaches, mostly initiated by nonstate actors.

Ten years later, the assessment of these approaches during Rio+20 was more than uncertain (Ramstein 2012). The voluntary commitments within the United Nations seemed to have encountered several limitations (see box 10.2) that transformed this "positive" sustainable

BOX 10.2 THE SPLENDOR AND MISERY OF VOLUNTARY COMMITMENTS

The Rio+20 voluntary commitments could have been a key legacy of the conference, which has otherwise been criticized for its low level of ambition and official outcomes. Presented as a complementary tool to international conferences aiming at enhancing sustainable efforts, these commitments were supposed to involve a wide audience of stakeholders, not limited to national governments, in a more participatory approach and in a large range of sectors. They were also meant to mobilize more funding needed to enforce sustainable strategies and policies.

Designed as such, however, voluntary commitments leave room for skepticism, if not fears, as to whether they can deliver on their promises. What is their level of ambition? How are they selected? Is their relevance and feasibility properly assessed? How will they be monitored and verified? Given their non–legally binding character, will governments use them to avoid responsibilities and stricter regulations?

According to the Rio+20 final outcome, a first step to ensure that promises are kept should consist of an aggregation of voluntary commitments and other commitments in a global registry. Acting as a pledge reminder inside and outside international conferences, this compilation would need to be based on regular, solid, pragmatic selection and reviewing processes. Thematic advisory boards, including different types of actors, could then be established to assess common indicators and progress on specific areas. The transparency and accessibility of this registry would allow bottom-up accountability, which would ensure the commitments' sincerity and avoid "green washing" initiatives (Ramstein 2012).

development agenda, based on voluntary action of multiple parties within a flexible and nonbinding framework, into a model confronted with the same difficulties faced by its intergovernmental predecessors: low levels of ambition and problems with implementation and evaluation.

On July 1, 2013, the UN Department of Economic and Social Affairs published a special report on the partnerships and initiatives for sustainable development that have been registered since Rio+20. It was estimated that the total amount of these multiactor contributions reached $600 billion. These commitments concern different issues: mainly education (with 328 initiatives), but also poverty reduction (304), maternal and child health (154), access to energy (140), and the Global Environment Facility (12). While welcoming these contributions, the report also recalled the fact that they can only complement the actions led by governments committed to in Rio (UN Department of Economic and Social Affairs 2013).

MORE AND MORE COMPLEX COOPERATION REGIMES

This chapter has emphasized that the particularity of global environmental governance lies in the fact that an inherent learning process is one of its major features. In this section, we will be more precise on the consequences of this fact: experimentation and processes of trial and error do not necessarily lead—aside from few exceptions (such as the replacement of the Commission on Sustainable Development with a high-level political forum)—to the disappearance of the poorest-performing instruments; rather, they tend to lead to their overlapping or integration with other layers of measures. Nothing actually disappears; instead, new initiatives emerge, are added to previous ones, and overload cooperation spaces.

The different coordination models that appeared over the past forty years in the field of international action for the environment did not substitute for each other. Rather, these models cohabitate, and different actors, such as governments, economic agents, NGOs, scientific communities, and subnational public authorities, make use of these platforms according to their strategic needs, convictions, and judgments on each instrument's respective appropriateness and efficiency. The coordination regimes, far from being simplified over time, have become more and more complex and interdependent. The complexity and interdependence are engendered by two elements: the increasing understanding of the interactions between environmental problems and human activities, and the proliferation of actors who feel concerned with it, whether through their perceptions of the impacts or their implication in the search for a solution.

There is one simple illustration of this argument. The communities involved in the analysis and management of marine ecosystem degradation (pollution, resource destruction, and ecosystem mutations) are organized in a specific manner and have sought to establish a sui generis ocean protection regime. The interaction with the carbon cycle and its impact on ocean acidification eventually forced these communities to cope with these interactions, building on the system dealing with climate change, which is already a multiform one. If global warming issues are addressed separately, the negative impact on marine ecosystems could be totally discounted, as is the case in some geoengineering solutions against climate change that promote the increase of carbon capture by oceans.

SCIENTIFIC KNOWLEDGE: A CORNERSTONE OF INTERNATIONAL COORDINATION

Scientific studies have driven the international community's reaction to and organization against environmental risks, as already observed several times in this book. Thus, science is one of the most important mainsprings of action.

To evaluate such environmental damages and risks as pollution, changes in climate and the environment, and management of marine resources and to make recommendations or come to agreement on evaluation mechanisms, negotiating parties indubitably need strong knowledge upon which they can base the definition of agendas, compromises, and validation and control mechanisms. The need for coordination in the production of scientific knowledge is strengthened by the fact that their "validation" by stakeholders is, in many cases, an issue of concern throughout the process of defining programs and policies. These productions may take different forms: from constituted epistemic communities, as in the case of the IPCC and climate change, to more or less formalized, coherent, and compatible networks. There has been a proliferation of panels, scientific instruments, environmental journals, reports, and evaluations by academies of sciences and others that alert societies and decision makers to the environmental damages caused by our economies—for example, the Millennium Ecosystem Assessment (MA) in 2005.[1] The MA has involved the work of more than 1,360 experts worldwide. Their main findings show that "over the past 50 years, humans have changed ecosystems more rapidly and extensively than in any comparable period of time in human history, largely to meet rapidly growing demands for food, fresh water, timber, fiber and fuel. This has resulted in a substantial and largely irreversible loss in the diversity of life on Earth."

Sixteen years earlier, U.S. political scientist Peter Haas (1989), in his studies on the Mediterranean Action Plan, a coordination framework for pollution control in the Mediterranean basin, demonstrated that "this regime played a key role in altering the balance of power within Mediterranean governments by empowering a group of experts, who then contributed to the development of convergent state policies in compliance with the regime. In turn, countries in which these new actors acquired channels to decision making became the strongest proponents of the regime" (51).

International networks for expertise and knowledge exchange play a key role in determining the causes and effects of highly complex problems, the support of states for the identification of their interests, the coordination of public debates, and the recommendation of implementation policies. Control over knowledge and information thus becomes an issue of power relations, and their diffusion can lead to the adoption of new behaviors. Haas (1992) further stated that these networks represent an epistemic community, which he defined as "a network of professionals with recognized expertise and competence in a particular domain and an authoritative claim to policy-relevant knowledge within that domain or issue-area" (21).

A MULTITUDE OF ACTORS

In addition to these knowledge communities, a range of actors is involved in environmental negotiations.

In Rio in 1992, the UN Conference on Environment and Development (UNCED) benefited from the numerous changes that took place during the 1980s. In fact, several actors gained a more informed understanding of the environmental question and invited themselves to the negotiation tables; these actors included nongovernmental or civil society actors, and later indigenous peoples, as well as local communities, which had practically had no place in international relations and multilateral organizations before. Today, just like state actors, "NGO diplomats have access to a number of resources that give them power in multilateral negotiations. Although they rarely possess significant military capabilities, some NGOs have considerable economic resources, particularly in the private sector" (Betsill and Corell 2007, 7). NGOs thus play a multifaceted role: "Activist groups provide research and policy advice, monitor the commitments of states, inform governments and the public about the actions of their own diplomats and those of negotiation partners, and give diplomats at international meetings direct feedback" (Biermann and Pattberg 2012, 14).

The number of NGOs represented at environmental megaconferences has continually increased since the Stockholm summit in 1972, with its 255 nonstate representatives: from 1,500 at the 1992 Earth Summit, the number of nonstate participants rose to 15,000 ten years later in Johannesburg, including several private-sector organizations. The latter followed closely the environmental negotiations. Industrialist Stephan Schmidheiny, principal author of *Changing Course: A Global Business Perspective on Development and the Environment* (1992), eventually inspired the creation of the World Business Council for Sustainable Development soon after the 1992 Rio Earth Summit. In fact,

> Business plays a key role in international environmental politics. Private firms are engaged, directly or indirectly, in the lion's share of the resource depletion, energy use, and hazardous emissions

> that generate environmental concerns. At the same time, firms can also serve as powerful engines of change, who could potentially redirect their substantial financial, technological, and organizational resources toward addressing environmental concerns. The environmental impact of firms' activities makes them central players in societal responses to environmental issues. (Levy and Newell 2005, 16)

At the time, the private sector—or at least a portion of it—understood the need for a change in its modes of operation. The initial Rio summit thus represented a moment of relative consensus among political representatives, NGOs, and the private sector on the fact that our consumption and production patterns are not compatible with environmental protection. Today, however, spaces for advocacy and political pressure are no longer the same. Until the end of the 1980s, NGOs were putting pressure on states to regulate global governance. After Rio, the link with businesses was established along a new type of relations that did not resort to state mediation and that gave rise to unprecedented partnerships. Today, although states are still in charge of building the international rules, they are more dependent on multilateral institutions, social movements, NGOs, scientific communities, and the private sector, which are all monitoring the implementation of public commitments and policies.

The dynamics generated by this multiplicity of actors have created new institutional arrangements and instruments of environmental governance. The implementation of norms for voluntary certification in environmentally sensitive areas, such as forests (see the Forest Stewardship Council created in 1993) and marine resources, or in the extractive sector (e.g., gold and diamond mining), are illustrations of such hybrid approaches. These norms make use of market incentives and consumer education to develop more sustainable modes of production.

This complexity ultimately results in a kind of meta-regime, as defined by Kate O'Neill (2007), an ecologist at University of California–Berkeley: "Perhaps the primary utility of viewing global environmental governance through the perspective of a meta-regime, consisting of rules, decision-making procedures, norms and principles, and actor roles is that it provides a dynamic, macro-level perspective on an evolving governance architecture, which nests the various individual international environmental treaty regimes" (12). The multiplicity of actors involved in environmental issues is greater than in any other field. It now includes different levels of collective action, different scales of space, and actors of different natures, as more and more actors progressively feel concerned. Indeed, some local authorities in France and the United States recently adopted GHG reduction targets that were even more ambitious than national ones. Governor Jerry Brown, who argued that "California is at the epicenter of climate change," even went further: after making battling climate change one of the state's priorities and adopting a forceful plan, he fervently tried to convince other leaders of states and countries to follow his lead (Jennifer Medina, "In California, Climate Issues Moved to Fore by Governor," *New York Times*, May 20, 2014).

The climate agreement in 2015, as well as several agreements in the environmental field today, includes a feature chapter recognizing the importance of regions and local authorities in the fight against climate change. These local actors are taking their share of responsibility for contributing to a global problem and are thus challenging the traditional division of tasks between nation-states—normally the only legitimate actors in the sphere of international relations—and local levels. These regional and local initiatives have the particular ability to remind states of their commitments, to profile solutions, and to finally contribute to making environmental governance less dependent on domination relations characteristic of the international diplomatic scene, which is still more familiar with issues related to the Treaty of Vienna than with the geophysical limits of our planet.

A "SECOND CLASS" ISSUE

The environmental question is a multiform, multiscalar, polysemous, intrinsically radical, long-term, and controversial issue; it also holds the "challenger position" on the international scene against more conventional diplomatic issues, such as security, the economy, and development. The history of the emergence and persistence of the environmental question in the international agenda is the history of struggles for power and domination, from the composition of ministries and departments to the bureaucracy staff.

A SHIFTING CENTER OF GRAVITY

In 2000, there was a shift in the balance between environmental and developmental issues. This shift came along with the agenda established by the Millennium Summit of the United Nations in New York and the Millennium Development Goals (MDGs) for poverty reduction, which marginalized environmental protection and sustainable development. According to Lerin and Tubiana (2005):

> This shift had numerous consequences. It endorsed the establishment of environmental protection as a secondary objective of the international community's agenda. Although the substantial environmental objective was not explicitly in dispute, its urgency and its hierarchical position on the international agenda, as well as the specific modalities by which it should be handled were clearly challenged. (82)

In 2008, scholars still wondered about this "crisis of the global environmental governance" and questioned,

> How and why have serious approaches to global sustainability come to be so politically marginalized? How is it possible—15 years after UNCED, 20 years after the Brundtland Commission, and more than 30 years after the Stockholm UN Conference on the Human Environment—that the great global challenge of securing the ecological future of the planet and its peoples has reached a point of such political and social insignificance? Much of the problem lies, unsurprisingly, in opposition from powerful interests. But the problem has significantly been worsened by an inadequate grasp of the linked challenges of sustainability, globalization, and governance. (Park, Conca, and Finger 2008, 5)

THE FINITE MEANS OF ENVIRONMENTAL GOVERNANCE

A whole range of international bureaucracies has developed to assemble the secretariats of multilateral environmental agreements. Although international bureaucracies seem to be overabundant, they are, in reality, poorly endowed. In fact, in 1995, the *New York Times* pointed out that the secretariat of the Ramsar Convention (the Convention on Wetlands of International Importance especially as Waterfowl Habitat, 1971) has a total of seventeen staff, including six professionals, to assist eighty-five parties; the CITES Secretariat has a total staff of twenty-two (fourteen professionals) to service 128 parties; the Basel Secretariat (Convention on the Control of Transboundary Movements of Hazardous Wastes and Their Disposal, 1989) has four professional personnel to deal with eighty-three parties; and the secretariat for the UN Framework Convention on Climate Change (1992) has nineteen professional staff and a total staff complement of thirty-one to assist 118 parties. Some have argued that it is only the considerable staff motivation that has ensured their effective functioning:

> Secretariats are the Cinderellas of international environmental treaty systems. They are often neglected, regularly criticized, and seldom rewarded for assisting governments in meeting their treaty implementation obligations. To a large extent, it is the satisfaction that personnel appear to get from their commitment to help implement a treaty that keeps many of them going under difficult conditions. Secretariats also appear to use their linchpin status, boundary spanning activities, networks, and professional expertise to significant effect in overcoming organizational constraints of limited authority and resources. (Sandford 1996, 11)

THE BURDEN OF PROOF

The scientific legitimation of environmental questions seems to be endless. An almost permanent process of evidence building around the few global environmental issues is at work, and scientific controversies are legion—most notably, those concerning climate. According to Miller and Edwards (2001), "For the emerging climate regime, continuing scientific controversy, and a dearth of simple solutions, render relations between expert knowledge and environmental governance even more important, and far more contested" (3).

This everlasting search for scientific justification prompted international development agencies to spell out their position on the subject in a call for action. The World Bank, for example, commissioned a scientific report in 2012, emphasizing that unless some concrete measures against climate change were implemented, the international community would have to face the catastrophic consequences of a 4°C rise in average annual temperature until the end of the century (extreme heat waves, decline in world food stocks, rising sea levels, etc.), which could affect millions of people (Potsdam Institute for Climate Impact Research and Climate Analytics 2012). This document, which collected the most recent scientific data on climate, indicated that the current GHG emission reduction commitments will not allow us to alleviate this increase in average temperature. Since then, all recent publications collecting new data, had been ever more pessimistic on the speed of global warming and the severity of impact.

This legitimation of the question of climate change by international institutions is at pains to conceal numerous—and sometimes violent—campaigns of scientific delegitimization, which use uncertainty as a pretext for justifying inaction when it comes to GHG emission reductions. This "environmental skepticism" goes back to 1998 and the first edition of Bjørn Lomborg's (2001) controversial book, *The Skeptical Environmentalist*, which postulated that "overpopulation, declining energy resources, deforestation, species loss, water shortages, certain aspects of global warming, and an assortment of other global environmental issues are unsupported by analysis of the relevant data" (6). Environmental protection regularly finds itself subject to the conjunction of hostile interests that weigh heavily in political arbitrations—especially in times of economic crisis, when the environment once again becomes a contingency external to the system it actually attempts to replace. Nevertheless, this regular attempt to put environment protection to the back burner is challenged by mounting evidence and public awareness.

AN OUTLOOK: THE WAY FORWARD

THE INTEGRATED AGENDA: PUTTING A PRICE ON ENVIRONMENTAL COSTS AND BENEFITS

The environmental community's preferred solution remains the internalization of externalities, although this has led to rather limited and disappointing results so far. Undeniably, initiatives and mechanisms for taxing CO_2 emissions exist—notably through taxes or emission trading systems—but they are still marginal. And they are far from sufficient given the growing dimension of the problem.

Following the same rationale of correcting market failures, efforts have been undertaken to uncover the economic value of services provided by the environment. One illustration of these efforts is The Economics of Ecosystems and Biodiversity (TEEB, 2008) initiative, led by Pavan Sukhdev, a former economist at Deutsche Bank, who asserted in an article in the *Guardian*:

> Holistic economics—or economics that recognise the value of nature's services and the costs of their loss—is needed to set the stage for a new "green economy." . . . The crisis of biodiversity loss can only begin to be addressed in earnest if the values of biodiversity and ecosystem services are fully recognised and represented in decision-making. This may reveal the true nature of the trade-offs being made. ("Putting a Value on Nature Could Set Scene for True Green Economy," February 10, 2010)

In the same vein, economist Nick Stern coordinated a report commissioned by the British government to quantify the costs of inaction on climate change. In an interview for *The Observer*, he stated that his study—an alert in 2012—underestimated these risks and costs: "The planet and the atmosphere seem to be absorbing less carbon than we expected, and emissions are rising pretty strongly. Some of the effects are coming through more quickly than we thought then" (Stewart and Elliott 2013). His 2006 *Stern Review* pointed to a 75 percent chance that global temperatures would rise by 2°C–3°C above the long-term average; in the interview, he explained that he came to believe that we are "on track for something like four." Had he known the way the situation would evolve, he said, "I think I would have been a bit more blunt. I would have been much more strong about the risks of a four- or five-degree rise."

The question raised by these market instruments is not whether this approach might lead to a "mercantilization" of nature, as denounced by some NGOs, but whether it will lead to one of the conditions of their implementation and success. The unsuccessful attempt to carry out a reform introducing a carbon tax in France in 2009, though well prepared by a commission of experts, economic actors, civil society, and union representatives, demonstrates how severely details can hamper such processes and how recommendations by economic actors can sometimes be ignorant of the application conditions of the measure they have been proposing. Although a governance model based on price mechanisms certainly can be economically efficient, the major drawback is that it can result in high transaction costs (considering the need for education, debates on the use of tax revenues, compensations, and eventually exemptions), especially when public opinion and economic operators are not convinced by the incentive virtues of its effects.

This is why the recent progress made in the promotion of carbon pricing by business networks and international institutions like the International Monetary Fund and World Bank are playing a decisive role in changing the perceptions at least of global economic actors and governments. The Carbon Price Leadership Coalition, launched by Christine Lagarde and Jim Yong Kim in September 2014, now involves seventy-four countries and more than a thousand companies promoting carbon pricing policies, exchanging experiences, and training officials. Major companies are now advocating a carbon price in the countries where they operate. If there is a progressive alignment of policies of this sort, new questions around the distributive aspects of these measures will have to be addressed. In addition, as carbon prices set in, different economies will stay differentiated, and, at least for a while, the competitiveness question will come back to the front.

FOSTERING CONVERGENCE IN EXPECTATIONS THROUGH MULTITIER STAKEHOLDERS' PARTICIPATION PROCESSES

In 1992, the Rio Declaration on Environment and Development confirmed its Principle 10 according to which "environmental issues are best handled with the participation of all concerned citizens, at the relevant level. . . . States shall facilitate and encourage public awareness and participation by making information widely available." Since 1992, direct forms of civil society participation—that is, the involvement of citizens or their representatives in the policy-making process— have started to emerge in the framework of international summits on sustainable development. Such mechanisms, including multistakeholder dialogues and Internet consultations, increasingly complement traditional forms of civil society participation, such as the Major Groups, which were established by Agenda 21 to channel the participation of citizens in international efforts to achieve sustainable development within the UN framework.

Such proliferation of civil society consultations, which are often organized and sponsored by governments and international organizations, suggests that direct participation of all relevant civil society actors in intergovernmental policy making is considered essential to the successful delivery of sustainable development policies. Consistent with this assumption is the statement made by UN Secretary-General Ban Ki-Moon: "The post-2015 development framework is likely to have the best development impact if it emerges from an inclusive, open and transparent process with multi-stakeholder participation" (UNSG report General assembly 2011).

However, whether or not civil society consultation processes are inclusive, open, and transparent is a subject of debate for two main reasons: one that questions the democratizing potential of civil society in global governance, and one that is skeptical about the representativeness of the consultation processes and the quality of policy outputs. Specifically, regarding civil society, the debate finds its origins in a recurrent but problematic assumption in the literature on global governance that conceives civil society as a force for democratizing the global order. While many authors acknowledge the constructive force of civil society in bringing expertise and voicing the interests of the affected and marginalized, they also warn against naive views of civil society organizations (CSOs) as representatives of the public good and as actors free from self-interest (Bäckstrand 2006). Indeed, civil society is not necessarily more inclusive, accountable, or representative than the market or the public sector (Scholte 2002). Likewise, the challenges faced by civil society can undermine the democratic legitimacy of both participatory processes and intergovernmental policy making.[2] Participatory processes, for instance, may reflect an unfair representation of civil society actors and may be permeated by the relative power of interest groups over the views of a broader public, due to an unavailability of resources and a lack of social mobilization (Downs 1957; Breyer 1993; Gastil and Levine 2005). Moreover, some scholars argue that participatory processes produce undesirable policy results at substantial costs, stressing that inclusiveness eventually hampers the effectiveness of sustainable development policies (National Research Council 2008). Likewise, others note that although civil society consultations produce recommendations that are passed on to heads of states or governments, they have, in practice, very little direct influence on the outcomes of intergovernmental negotiations on sustainable development (Rask, Worthington, and Lammi 2012).

Against this backdrop, several initiatives have emerged that could bridge the gaps mentioned above. The Sustainable Development Dialogues (SDDs) were organized in the framework of Rio+20 by the government of Brazil with the support of UNDP; they engaged more than 60,000 people from 193 countries. The goal of the dialogues, which consisted of two phases, was to foster debate among stakeholders and citizens on ten topics. The SDDs were first launched through a digital platform[3] to provide individuals with a space for discussion. Facilitated by academic experts, the online discussions on the ten themes resulted in a set of recommendations, for which the online participants voted through the platform's voting system. These recommendations were then transmitted to the participants present at the conference venue in Rio Centro. During the second phase, top representatives from civil society engaged in "open and action-oriented debates" at the conference venue to agree on thirty recommendations—three for each topic—from the series provided by the online consultations. This final set of recommendations was directly conveyed to the heads of state and government present at the high-level segment of Rio+20 and was added to the annexes of the Report of the Conference. In addition, each set of three recommendations was submitted to the vote of the wider public.[4] Further connecting these initiatives to the broader public opinion debate and to the more restricted circle of negotiators and officials is the way forward after what has been experimented with in Rio.

UNLOCKING PRIVATE FINANCE

The Green Growth Action Alliance (G2A2), a multistakeholder coalition including companies, public and private financial institutions, and research organizations that was launched during the 2012 G20 Summit in Mexico, published its first report upon the occasion of the 2013 World Economic Forum. The study, which established climate change as a systemic risk for the world economy, explores ways and means to support the financing of adaptation and mitigation measures. In 2012, climate change cost the world economy $1,200 billion, or 1.6 percent of the world's gross domestic product (GDP) and directly resulted in the death of five million people. In 2030, this cost could reach 3.2 percent of the world's GDP and swallow up to 11 percent of the least developed countries' GDPs. According to the study, supporting sustainable economic growth that respects the targets for climate change mitigation would require $700 billion of investments on the global scale. Institutional investors should lead by example and increase their contributions by $36 billion, which would make the worldwide public budget reach $96 billion. The consequences of such an engagement would not only be limited to this increase of volume but would also create the required investment conditions by providing political risk insurance, ensuring the stabilization of energy and climate policies in developing countries, and so forth. This reinforcement would offer strong incentives to private investors, whose contribution could reach up to $570 billion, according to the study (Green Growth Action Alliance 2013).

Since the publication of the G2A2 study, many more works have been launched in an attempt to design the new financial ecosystem needed to deploy the Sustainable Development Goals and sound climate policies. Consultations and study groups are flourishing, in particular around the G20 but also in many informal settings. This work in progress may lead to systemic change in the future.

CREATING A FRAMEWORK FOR EXPERIMENTATION AND ACCOUNTABILITY AROUND THE SUSTAINABLE DEVELOPMENT GOALS

The negotiation of the Sustainable Development Goals (SDGs) was one of the key outcomes of Rio+20. As this chapter has shown, after a retrospective analysis of the achievements and limitations of the first Rio conference in 1992, the SDGs *should* help address two challenges: the implementation and the coherence of sustainable development policies in all countries, regardless of their income level. The SDGs thus need to embed sustainable development into economic reality; encourage learning and experimentation processes, given the lack of "solutions" and recipes; and, in so doing, restore the virtues of cooperation. The conditions for success are nevertheless demanding.

The translation of sustainable development into economic reality, which follows on from its translation into political reality, creates losers; its added value is crippled with uncertainty, as uncertainty about the benefits may exceed uncertainty about the costs of action. Making sustainable development operational requires the creation of an *internal* political compromise in each country or region and the rejection of the overly simplistic idea of win-win solutions. A compromise between countries made the concept of sustainable development viable in 1992; the compromise reached in 2015 within countries makes it operational, as the negotiations were not limited to deciding what is good for others—especially the developing and least-developed countries—but have been instead an opportunity to answer the eminently less consensual question of what is good for oneself.

Making sustainable development operational also requires a comprehensive and coherent accountability framework whereby any actor involved or concerned could learn from others and test innovative solutions. The 2030 agenda and the SDGs could provide all of this. Challenges and perspectives are clear; what's left to be done is to make the 2030 agenda and its SDGs deliver on them. The good news is that many stakeholders and governments are now trying to study what it takes to implement SDGs in practical terms and begin to report on achievements.

11

THE GEOPOLITICS OF ENVIRONMENT

A G2 MODEL BETWEEN THE UNITED STATES AND CHINA

As illustrated in chapter 10, the global environmental question is a fundamental question of international relations, while simultaneously being regularly sidelined within the global agenda. As such, environmental governance is inherently linked to the question of sovereignty: although global environmental governance has been reappropriated by intrastate levels (local authorities such as cities, but also regions and federated states) and by nonstate players from influence networks and actors of participatory democracy (such as civil society, associations, NGOs, and businesses), it persists in being dominated by the state. Indeed, the state remains preeminent in the framing and shaping of collective action (Lerin and Tubiana 2005; Tubiana and Voituriez 2007). The presuppositions from the 1990s—that is, that state levels would be marginalized in the future regulation of global issues—are disclaimed today by the resistance (if not the reaffirmation) of the ultimate sovereignty of states in deciding their own development path.[1] The persistence of states can be directly explained by the duality of international modes of collective action legitimation—between a legal order founded by states and a common culture that tends to transcend them.

Over the past twenty years, governments have largely been invested in this sphere of international relations, as reflected by Angela Carpenter's (2012) analysis: "While the first conference in Rio was attended mainly by scientists, attendance at subsequent conferences included representatives of environmental ministries from countries across the globe, together with representatives of non-governmental organizations (NGOs), and have seen shifting alliances between nation states with their own interests and negotiating positions" (163).

In this chapter and the next, our investigation of the geopolitics of global environmental governance will describe the classical relations of power and interests among states during the construction of collective action—that is, the most powerful and best-organized states tend to impose their agendas on others in order to promote their own interests. These relations must be distinguished from global environmental governance per se, the object of which is to define institutions (or the regime complex) for the management of global public goods, against which its performance can be measured.

The collective norms and standards implemented to resolve, mitigate, or compensate the effects of the global environmental crisis are the joint outcome of the power and interests of nation-states in interaction with the body of knowledge produced by the many communities involved in sustainable development. As we will see, the "resistance" of the state—especially after the realist backlash mentioned in chapter 10, which can be symbolically dated to the 2002 Monterrey Conference—has led countries to consider certain environmental services as global public goods or assets, while at the same time limiting or reducing efforts to protect them (instead of creating the necessary condition for their treatment in a global and efficient manner).

In chapters 11 and 12, we attempt to go beyond this apparent paradox through a geopolitical analysis of four specific cases:

- The first case is the position of the United States, the historical leader of the movement toward a multilateral regime, which later became the spearhead of the relegitimation of the defense of national interests.

- The second is the situation in China. That nation's tremendous economic growth and its international internal effects are transforming the country's negotiation position from a Third World–like "wait-and-see policy" to an increasingly involved and innovative one. In this respect, it might very well embody some kind of pioneering position for emerging countries in international negotiations, which can expand the options available to "third countries."
- The third is the attitude of the European Union, which indisputably contributed to the promotion of supranational and constraining commitments in the name of protecting global public goods—at least until the UN Climate Change Conference in 2009. The EU's engagement has been weakened not only within the community but also at the member-state level.
- The final case involves an analysis of the role of newcomers to the international arena—that is, the middle-income countries, which see the environment as a positive lever to increase their influence.

Chapter 11 focuses on the two major players: the United States and China. These nations are the two main emitters of greenhouse gases and the major consumers of natural resources. They are also the heavyweights on the international scene, and in many cases, they define the balance and center of gravity of any agreement.

China is becoming more and more assertive in its views about the global governance of the environment and wants to be on an equal footing with the United States. The United States is concerned with the potential impacts of environmental regulations on its competitiveness—in particular, vis-à-vis the former. Thus, there is a logical move to seek some form of coordination between the two. Is the setting of a G2 in global environment governance probable or possible? This chapter analyzes the political economy of the international environment policies of both countries and assesses convergences and divergences, with an understanding that recent developments in U.S. politics challenge previous (successful) efforts during the Obama administration.

THE PROGRESSIVE EROSION OF U.S. LEADERSHIP

The foreign policy of the United States and its doctrines are located within a space principally defined along two axes: the two extremes of the first axis are isolationism and interventionism, while those of the second are unilateralism and multilateralism. International coordination, beyond the balance of power and the Westphalian sovereignty model, was initiated during the interwar years with the League of Nations project, which the United States both initially promoted and eventually ruined. After the United States entered the war against Japanese imperialism and Nazi totalitarianism, it de facto started coordinating an international order proceeding via inclusion and exclusion—first of communist countries and later of the nonaligned states. For all these countries, the United States defined the multilateral system of world coordination and, at the same time, established rules and modalities for its own bloc. In the context of the battle of ideas, the ambition was not only to define solutions for one side but also to define outlines for solutions that are valid worldwide. These universalist ambitions were obviously obstructed by the fact that the interests of the two blocs were divergent and even contradictory. The U.S. position never succeeded totally in liberating its formulation of universality from its opposition to a countermodel.

The environmental question is no exception to this dualism, which is an essential part of U.S. leadership. American hegemony has always been partial and weakened by the fact that the U.S. model had to internally and intrinsically prove its efficiency, which ultimately was brought about, by default, by the collapse of the Soviet system and, more generally, of communist regimes at the end of the twentieth century. The handling of environmental issues at the international level did not escape this evolution of the definition of the role and interest of the United States. Global, multilateral solutions certainly have to demonstrate their efficiency at domestic levels first. By the same pragmatic reasoning, to ensure their implementation and success, global and universal arrangements must either demonstrate their worth or move forward along a fragmented agenda.

The sectoral and institutional approach characterizing the treatment of environmental issues is directly inspired by the U.S. administration's operation mode: "Since the 1970s, global environmental issues have been dealt with in a compartmentalized way by negotiating issue-specific treaties and building institutions around them" (Held 2013, 104). The universalist ambitions of the United States are, on the one hand, strongly tinged with pragmatism and, on the other, influenced by exceptionalism. In substance, they affirm the idea that what is good for the United States is good for the world—or, more precisely, that it is within the United States that one will find the innovative solutions to global problems.

The ambiguity of the current situation of global environmental governance certainly relies on the fact that the United States has always sought hegemony based on the exemplarity of their model (hegemony being an effect, not a cause). At the same time, this ambiguity stems from the diffusion of U.S. domestic policies and models to conceptualize its own vision of universality. By nature, this situation does not facilitate negotiations. Obtaining U.S. participation, however, has often been a guarantee of the efficiency and success of—if not a necessity for—any international environmental regime. As a result, the successive visions developed by the U.S. administration have spread across all international negotiations, guiding the form and scope of agreements, regardless of the willingness of the United States to participate.

AN ILLUSTRATION WITH CLIMATE POLICY

The Kyoto Protocol of 1997 can be considered the last environmental agreement negotiated with the United States acting under the model of a constraining universal agreement. In fact, the protocol "is intended to be universal in its application, applying to all countries according to agreed principles of burden sharing; it is universal in its negotiation and decision-making process, being based on the primacy of the UN framework; and it seeks to establish legally binding international obligations." The protocol also proposed a broad approach to the environmental question, with the ambition of responding to the full range of problems provoked by climate change, as "it prescribes, in a top-down way, generally applicable policies that are based on commonly understood principles" and "strives to develop targets and instruments of climate governance (regarding mitigation measures, carbon sinks, adaptation efforts) in a comprehensive manner" (Falkner, Stephan, and Vogler 2010, 256–257). It thus set objectives and targets for GHG emission reduction within a specific timeframe. To organize mitigation efforts, it created a market mechanism for emission regulation, replicating the CO_2 emission trading rights in the United States, as opposed to the European idea of a carbon tax. As we know, however, the United States never ratified the protocol.

American reservations have been perceptible since the first days of the establishment of a climate agenda. Indeed, Washington only reluctantly signed the Rio Convention on Climate Change in 1992: President George H. W. Bush traveled to Rio to sign the UNFCC "only hesitatingly and under pressure, . . . agreeing to the treaty because there were no binding limits placed on the United States" (Roberts 2011, 777).

The Clinton administration (1993–2001), with Al Gore's participation, was more open to this agenda and took part in the negotiation of the protocol until its signing in 1997 in Kyoto. The administration, nonetheless, took care to ensure a "burden sharing" restricted not only to developed countries, insisting on the need to define emission constraints for emerging countries, such as India and China, as well. Despite a number of concessions by other parties, the Clinton administration did not succeed in obtaining ratification by the U.S. Congress due to the Byrd-Hagel resolution (see chapter 9).

In 2001, George W. Bush, the newly elected president of the United States, denounced the political commitment taken by his predecessor and announced that the United States was withdrawing from the Kyoto Protocol, despite protests from the international community. He also worked to reduce the legacy of Rio, whose tenth anniversary was being prepared: "The new administration set up a new position and was very active in reshaping the agenda of Johannesburg Summit" (Lerin and Tubiana 2005, 81). The White House blocked any agenda that would comprise an assessment of progress ten years after Rio. Instead, it supported the launch of public-private partnership "initiatives" that would bring private-sector contributions alongside the action of public institutions to promote sustainable development. This resulted in a new mobilization of the private sector; ironically, that mobilization would lead, thirteen years later, to an active campaign by business leaders in favor of an ambitious climate change agreement.

The Bush administration declared that it intended to search for a more equitable sharing of the burden of international environmental protection—that is, it asked that emerging countries, especially China, accept the same constraints as the United States. The United States thus developed a peculiar understanding of the principle of "common but differentiated responsibilities" enshrined in the UNFCCC. The United States interpreted this quest for "burden sharing" as a protection of national interests, leading it to try to impose unilateral positions whenever possible. The U.S. administration's intent was to promote technological—or at least unbinding—reduction instruments, while at the same time minimizing the delegation of power to a cooperative system.

This position ultimately isolated the United States in UN climate negotiations. In 2005, the Kyoto Protocol entered into force, after finally reaching the condition stating that at least fifty-five parties to the convention, accounting for 55 percent of the total 1990 CO_2 emissions by developed countries, had to ratify the treaty. They reached this position thanks to Russia's ratification, as negotiated by the European Union.

The U.S. position of defending its national interests resulted in a situation in which the country was considered uncooperative and isolationist in addressing a global issue. To overcome this damaging reputation, the Bush administration launched several counterattacks, though they did not succeed in completely delegitimizing UN negotiations.

In July 2005, the Bush administration tried to create a parallel track through a new partnership: the New Asia-Pacific Partnership on Clean Development and Climate included, alongside the United States, Australia, China, India, Japan, and the Republic of Korea. The partnership was intended to promote economic development and poverty alleviation and the development and diffusion of new energy technologies. This partnership did not go beyond joint statements without any operational deliverables.

The second initiative was somehow more successful and was intended to create an alternative setting to UN discussions, a club including a limited number of key countries. In 2007, a few weeks before the Bali Conference, at the invitation of the U.S. government, a Major Economies Forum convening the sixteen main GHG-emitting countries, including eight emerging countries, met to draw up a joint declaration on clean technology and climate change. This event did not get a lot of exposure. It did not reach the outcome the Bush administration hoped for—the weakening of the multilateral process—as the process of negotiation within the United Nations continued, but it did create a useful platform for informal exchanges. It has survived under different names until today.

The election of Barack Obama in November 2008 allowed the reopening of some environmental debates—most notably, on energy. The Democratic president redefined the relation of national security, energy, and the environment during his first inaugural address on January 20, 2009: "Each day brings further evidence that the ways we use energy strengthen our adversaries and threaten our planet."[2] This orientation was supported by the administration's domestic investments program: $42 billion out of a total of $78 billion was allocated to the development of renewable energy technologies, and $570 million was allocated to research projects on climate change (Pew Center on Global Climate Change 2009).

This orientation also encouraged the continuation of earlier movements for local GHG emission regulation, such as California's strategies to reduce emissions and fight climate change. The city and county of San Francisco, for instance, had launched its own climate action plan in 2002, which included measures fostering the development of renewable energies, accelerating the deployment of recycling projects, and improving transportation efficiency. In 2006, this plan expanded to the state as a whole, and California adopted a law that included several measures for GHG emissions reduction, together with a carbon market and emission norms. In 2014, a plan to reduce GHG emissions to 80 percent below 1990 levels by 2050 was adopted. In September 2015, Governor Jerry Brown promoted a coalition of states and Canadian provinces aimed at limiting GHG emissions to two tons per capita by 2050.

By 2015, thirty-three out of fifty U.S. states were applying renewable portfolio standards, which are systems enforcing the production of renewable energies. Furthermore, several regional alliances of states have been created to organize common systems of GHG emission reduction. For example, since 2005, the Regional Greenhouse Gas Initiative has convened ten northeastern U.S. states in a compliance carbon market. The Western Climate Initiative, a mirror of that initiative for the West Coast, also includes Canadian provinces (Schiavo 2011).

During his second term, Obama decided to accelerate this movement. On June 17, 2013, he launched a climate action plan focused on energy efficiency:

> The plan looks to reduce harmful greenhouse gas emissions in a comprehensive way and takes on the question of how to protect the country from the devastating climate-related impacts we are already seeing today. With a clear, national strategy in place—and concrete steps to implement it—the administration can protect people at home and encourage greater ambition internationally. Importantly, the president is recommitting the United States to meet its target of reducing greenhouse gas emissions by 17 percent below 2005 levels by 2020. (Morgan and Kennedy 2013)

On August 22, 2013, California announced that it had sold all its CO_2 emission permits for the first time:

> Starting in 2013, businesses that are major carbon polluters in California will need to turn in one of these allowances for every ton of carbon pollution they produce. California established a fixed supply of these permits in order to "cap" industry's greenhouse gas emissions, with the "allowance budget" established out to 2020. Because the cap declines each year, California businesses will need to reduce their pollution over time. (Kennedy 2013)

Despite these notable domestic trends, the U.S. administration's modus operandi in international negotiations did not really change, especially because of the persistent resistance of Congress to all external engagement that could endanger the economic competitiveness of the United States:

> The President, although in charge of foreign environmental policy, faces two powerful constraints at the domestic level: the political coalitions in Congress that determine the pace and direction of domestic legislation on climate change; and the efforts by interest groups—business and NGOs—to shape events in Congress and on the international scene. In other words, U.S. climate policy flows from the domestic to the international, and global environmental leadership that is credible requires, first and foremost, strong domestic action. (Falkner 2010, 38)

The United States continued to favor a segmented approach to environmental problems and a definition of issues that was as narrow as possible, in opposition to a concept of one broad, deep, and binding agreement. This approach of the United States advocated for a sectoral and technological treatment of problems, as the two following examples illustrate.

THE CASE OF MERCURY: LOOKING FOR FLEXIBILITY

The issue of mercury, a neurotoxin inducing damage to humans and the environment, has been on the international agenda for several decades: "Mercury pollution has been subject to high-level domestic political concern since at least the 1950s, and continuing international policy making since the 1970s. Levels of mercury in the atmosphere nevertheless have increased by a factor of three since the beginning of the industrial era" (Selin and Selin 2006, 258).

Until the 1990s, Europe and the United States were the main mercury consumers. The international mobilization around the issue progressed slowly. UNEP used its expertise to try to turn it into an issue that could not be ignored; it eventually succeeded in the early 2000s: "The process towards a global treaty began with a scientific assessment report, the 2002 Global Mercury Assessment. A main conclusion of that assessment was that there was sufficient evidence of significant global adverse impacts to warrant international action to reduce the risks to human health and/or the environment arising from the release of mercury into the environment" (Selin 2013). Three possible courses of action were then put on the table:

- The creation of a global convention was rejected by countries exploiting heavy metals—the United States, Canada, Australia, and New Zealand—and their major consumers—China and India.
- A regulation under the auspices of the Stockholm Convention on Persistent Organic Pollutants was not politically plausible because the United States did not ratify it, even though 127 countries and the European Union did. It thus would have lacked both legitimacy and efficiency.
- The development of voluntary partnerships appeared to be the least costly solution, because it involved producers and did not require national legislation (Selin and Selin 2006).

The third option was pursued by the United States in 2003 and finally used in February 2005 during the twenty-third session of the UNEP Governing Council and the Global Ministerial Environment Forum (UN Environment Programme 2005). Under this initiative, the idea of an international treaty with specific targets for emission reduction and the elimination of certain products was developed with support from the European Union and the United States. In February 2009, negotiations for a binding instrument were launched, ending on October 10, 2013, with the signature of a treaty in Minamata, Japan—a city sadly notorious for an environmental disaster involving mercury poisoning in the mid-twentieth century. The United States was the first country to sign and ratify the Minamata Convention.

This convention has not been free of criticisms, most of which focus on two aspects: (1) the overly long time allocated for the inventory of emissions (about five years), even though available scientific expertise would already have provided sufficient data for the determination of realistic reduction targets; and (2) the fact that reduction objectives and defined monitoring processes are still to be set. The treaty's philosophy is to ban primary mercury mining and mercury use in products where safe alternatives are already available; this organizing principle is clearly inspired by the Montreal Protocol, with the definition of "available alternatives" being, as expected, a matter of tough negotiations.

THE SINO-AMERICAN AGREEMENT ON HYDROFLUOROCARBONS

In Rancho Mirage, California, on June 8, 2013, the presidents of the United States and China signed an agreement aimed at the progressive elimination of hydrofluorocarbon (HFC) production and use. These gases, used for refrigeration, air conditioning, and industrial applications, are, on average, 2,800 times more potent than carbon dioxide in terms of global warming. Their use expanded after the phasing out of chlorofluorocarbons (CFCs) and hydrochlorofluorocarbons (HCFCs) under the Montreal Protocol in 1987, which was intended to protect the ozone layer. Although HFCs only represent 2 percent of today's global GHG emissions, they could reach up to 20 percent by 2050 if nothing is done to hamper their development. In Asia, where HFC emissions grow by 10 percent each year, it has become a key issue. China is willing to convince India and Brazil—the two major producers of HFC gases—to join this commitment.

In addition, the European Union announced its intention to reduce its fluorinated gas emissions by two-thirds and the total amount of HFC that can be sold within the EU by 80 percent by 2030. A global limitation of HFC gases could potentially reduce by 90 GT (gigatons) the CO_2 equivalent emissions until 2050, which represents approximately two years of all GHG emissions.

The Sino-American initiative is an illustration of the pragmatism and realism shared by these two countries on environmental issues. It also illustrates America's willingness to circumscribe problems in order to make them technologically "able to be fixed" and economically tolerable, with or without the UNFCCC's endorsement.

In the aftermath and enthusiasm of the climate agreement in Paris in December 2015, the U.S.-China initiative has given a push to the negotiation of an amendment of the Montreal Protocol to include HFCs, allowing a "docking station" to the bilateral (voluntary) agreement. After months of intensive diplomatic work, in particular by U.S.

officials, an amendment was approved in Kigali in October 2016 offering financial support to developing countries by an unprecedented new coalition of governments and philanthropic foundations. Like in Paris, the bilateral agreement process has finally been reconnected to the multilateral process.

The reluctance of the United States to engage in multilateral processes designing global frameworks, as well as its preference for fragmented instruments and bottom-up processes relying largely on voluntary pledges, has had a large influence on the recent evolution of global environment governance. One can draw diametrically opposed conclusions from these processes, depending on one's focus when analyzing its results: a focus on the substantial issue (that is, the need for adequate action up to the problem dimension) versus a focus on the existing framework and its constraints (notably, the divergent interests of international actors).

In the first perspective, the compromises reached do not meet the magnitude of environmental risks, because they give too much room to economic and national interests, ultimately leading to watered-down compromises. In the second perspective, which acknowledges the world and interests as they are, U.S. pragmatism and realism might open the way to a process of consultation, dialogue, and learning between parties meeting regularly and under diverse configurations. Because the pragmatism, the reliance on technological solutions, and the preference for bottom-up agreements that don't infringe on national sovereignty find a positive echo in China's international politics, they will consequently gain attraction globally. What remains key is the ability to "connect" these ad hoc processes with the broader frameworks in order to build trust between parties, change the positions of reluctant countries, obtain a consented and applicable agreement, and progressively increase the level of expectations and ambitions of international goals.

THE MUTATION OF CHINA

At the Seventeenth National Congress of the Communist Party of China, in October 2007, Chinese president Hu Jintao revealed a new doctrine for the country's future development—the "ecological civilization," meaning the addition of the ecological dimension to the nation's economic, political, social, and cultural development goals. This was a big shift from China's traditional stance, such as the 2002 slogan aiming to attain "four Chinas" by 2020, which led to the catastrophic situation of China's environment.

Since then, at the Eighteenth Congress in 2012, China's leadership emphasized the problems of uncontrolled economic development and environmental degradation. This evolution has major consequences both internally and internationally, as it has led China to redefine its traditional position from a country opposing environmental constraints in the name of development and poverty alleviation to new grounds that it is beginning to explore.

AN OVERCONSUMING ECONOMIC MODEL

China achieved a fascinating economic catch-up during the 1990s, which has slowed down only very recently. Its total GDP in 2012 accounted for twenty-four times its GDP in 1978, when China opened its economy (China economic information network statistics database, http:// db.cei.gov.cn). Since 1998, Chinese manufacturing production has increased fivefold, while production in Europe and the United States practically stagnated. China has thus become, according to World Bank figures, the world's leading exporter of goods, the world's largest importer of primary commodities, and the second-largest national economy in the world after the United States (World Bank 2012), if you don't count the EU as one single area, which would then make it a first-world economy.

With this economic catch-up and its spectacular industrial performances has come the most profound internal upheaval, most notably in spatial demography and urbanization. In 2013, 53.7 percent of the Chinese population lived in cities, up from a mere 17.9 percent in 1978. Between ten and sixteen million new inhabitants are expected to arrive every year in Chinese cities at least until 2030.

The result of such urbanization is not only the emergence of an immense internal market for goods and services; it also requires an intense mobilization of natural resources. In 2012, Chinese consumption represented between 40 and 50 percent of the global demand for natural resources. The Chinese consumption of raw materials and environmental resources corresponds to a double demand: the indirect demand by countries importing Chinese exports and the country's internal demand. Its low-priced goods have reinforced mass consumption in developed countries, which de facto outsourced the negative externalities of industrial mass production. The impacts of these externalities—in terms of various forms of pollution—are now dramatically being felt by China.

THE HIGH PRICE OF ENVIRONMENTAL DEGRADATION

The first concern about this growth, which emerged in public opinion and then in the Chinese leadership, was the local consequence of environmental degradation—mainly, air and water pollution. In July 2010, China's Ministry of Environmental Protection revealed that the number of environmental accidents almost doubled during the first semester and that the air quality of the biggest cities in the country had deteriorated for the first time in five years. It also stated that a quarter of the major rivers were dangerously polluted and that 180 cities were seriously affected by acid rain. In November 2010, the ministry recognized that the situation would continue to deteriorate, citing, for instance, the increase in the number of smog days in the biggest cities due to the explosion of car traffic (Jing 2010).

Since then the situation has worsened. In 2012, less than 1 percent of China's cities met the WHO air quality standards. In 2013 and 2014, pollution peaked in Beijing, Harbin, and Shanghai, where the level of hazardous particles in the air (in particular, in January 2013) amounted to forty times WHO norms. This generated citizens' discontent, as they dubbed the situation an "airpocalypse." Beijing's levels of PM2.5s (particles that are smaller than 2.5 μm [micrometers] in diameter and that can penetrate the gas exchange regions of the lungs) are the worst in the world.[3]

The Chinese media regularly publishes impacts on health, and several studies point out the economic costs of environmental degradation. For example, studies published by the U.S. National Academy of Science evaluate the loss of life expectancy in polluted areas to 5.5 years of life due to environmental degradation and an increase of diseases (such as cancer, pneumonia, and diarrhea), which can be now attributed to toxic emission from industries (Pope and Arden 2013). Some villages in particularly polluted areas are named "cancer villages" by Chinese social media (Jing 2010).

The economic costs are already being felt—especially in fisheries, as polluted rivers affect both aquaculture and catches in Chinese waters, and in agriculture, as water pollution affects the quality of the food produced. The World Bank estimates losses due to environmental degradation at 9 percent of the GDP (Economy 2007). Most experts, including those from government institutes, express serious doubts about the possibility that China's economy will keep growing as fast given the negative environmental externalities.

The second dimension of concerns about China's rapid growth is the impact on the global environment, as China is now the greatest GHG emitter. If trends continue as they are, the total amount of CO_2 emitted for energy production and consumption between 1990 and 2050 will be equivalent to the entire global emissions from the Industrial Revolution to 1970. China's economy has become the major driver of global warming, even though the emissions per capita are lower than those of many industrialized countries (though they are already higher than the per capita emissions of some European countries). It should be noted that China's per capita emissions are much higher than the world average. The Chinese government's new stance on climate change derives from this reality: without controlling its own emissions, China is shooting itself in the foot. Indeed, several reports published by the U.S. National Academy of Science (National Assessments of Impacts of Climate Change, 2007, 2011, 2016, cited in Gao Yun 2016, 236–237) state how much China is already affected by global warming through desertification, sea level rise, and disruption in water cycles, including flood and drought.

The impact on the Tibetan region is of utmost importance to China, as melting glaciers produce large-scale flooding and mudslides and reduce the runoff to the Yangtze and Yellow Rivers. The natural runoff volume of the Yangtze River is likely to decrease by 25 percent in the future—an accelerated phenomenon in that Tibet's surface temperature is rising faster than the global average.

Monitoring of sea level rise shows the same pattern. Impacts on China's coastline are above the global average. Shanghai and Tianjin, for instance, show higher rates of sea level rise due to their high concentration of buildings and their overextraction of groundwater, which decreases land load-carrying capacity and accelerates land sinking. The China Meteorological Administration predicts that changing climate trends on land in China include increasingly frequent and severe dust storms in northern China, less rainfall in eastern coastal areas, and stronger snowstorms in southern and central China. The most notable climate trend in the coastal areas includes increases in sea surface temperatures, resulting in more powerful tropical cyclones.

A RACE AGAINST TIME

The local and global environmental threats in China have generated a number of policy responses, especially because of constant pressure from China's civil society and citizen's protest movements, which are increasingly affected and concerned—if not exasperated—by the country's environmental situation:

The number of protests involving more than 100 persons amounted to 87,000 in 2005. The reason behind these social movements were recorded as being related to dispossession of lands and growing pollution. Since 2002 the number of complaints about environmental problems increased by 30 percent. The number of massive protests has grown by 29 percent each year, and in 2012 and 2013 large-scale protestation movements against industrial projects took place. In 2005, during the first mass demonstration in Dongyang, protesters claimed damages for the chemical pollution. Since 2012, demonstrations taking place are increasingly preventive, as in Xiamen or Shifang, or more recently in Zhongtai.

The National Environmental Protection Agency (NEPA) was created in 1988 under the supervision of the State Council. It was placed in charge of all environmental legislation, elaboration of national standards, management of pollution problems, and coordination with the United Nations. However, it did not get real power or adequate staffing levels. After an institutional reform, it was elevated to the rank of a super ministry in 2008. Moreover, the environmental targets were introduced in the decision making.

China's Eleventh Five-Year Plan (2006–2011) was the first to elaborate on a comprehensive environmental policy with binding targets. It aimed at reducing polluting emissions of SO_2 and diesel OC by 10 percent, as well as decreasing by 20 percent the energy intensity as compared with 2005 levels. This was not an easy task, as the government was forced to take exceptional measures to reach the goals. A program focusing on getting the 1,000 largest companies to cut their energy consumption by 100 million tce (tons of coal equivalent) was established using direct-control measures on the mainly state-owned companies. In an effort to comply with the plan, local governments decided to cut electricity supplies, forcing plants to use diesel generators, which actually caused an increase of polluting emissions and oil shortages. The National Development and Reform Commission's (NDRC's) vice minister nonetheless announced that, at the end of the Eleventh Plan, energy savings would exceed 600 million tce, which is the equivalent of 1.5 billion tons of CO_2.

The government also made an enormous investment in non–fossil fuel energy, making China the largest investor in renewable and nuclear energy. According to official Chinese figures, in 2011, the installed generating capacity of hydropower reached 230 million kW, ranking it first in the world. Fifteen nuclear power units were put into operation, with a total installed capacity of 12.54 million kW. Another twenty-six units, still under construction, were designed with a total installed capacity of 29.24 million kW, making China the world leader in nuclear energy. The installed generating capacity of wind power connected with the country's power grids reached 47 million kW, ranking it top in the world. Photovoltaic power generation also reported speedy growth, with a total installed capacity of 3 million kW. The state also expedites the use of biogas, geothermal energy, tidal energy, and other renewable energy resources.

Before completion of the Eleventh Plan, China had committed to reduce its growth of carbon intensity by 40 to 45 percent by 2020 as compared with 2005 levels. This was part of its pledge within the Cancun agreement in 2010. This pledge was translated into a law regulating the struggle against climate change, the major objectives of which are outlined by the Twelfth Plan. Furthermore, the national clean energy development plan (that is, development of non–fossil fuel energies) foresees bringing China's share of energy from clean power sources up to 15 percent of the nation's primary energy demand by 2020 with an investment of more than $360 billion.

Clean energy actually benefited from an extraordinary upturn in investment, supported by strong incentives: the 2006 law requires distribution companies to buy the electricity produced by solar power plants or wind farms and to spread additional costs across consumers. Producers with an installed capacity of more than 5 GW were obliged to produce at least 3 percent of their electricity from clean energies in 2010, increasing to 8 percent by 2020. However, in some provinces 40 percent of power from wind farms could not be transferred to the electric network in 2016, and in this same year 21 percent of the electricity produced by wind turbines was rejected by the grid (Zhang, Tang, Niu, and Diu 2016).

Seven pilot carbon markets were launched in China, and there were plans to start a national market already in 2016 but plans have been delayed. This national market, which should regulate about twice as many tons of CO_2 (3–4 billion tons of CO_2) by 2020, is similar to ones already in place in the European Union. It could double the value of the global market (Reuters 2014).

Finally, for the first time, the Twelfth Five-Year Plan (2011–2015) included targets for the reduction of nitrogen oxide emissions, which are responsible for acid rain, and of ammonia nitrogen emissions, which are blamed for their influence on water eutrophication processes. According to the State Council, €345 billion had been to be invested in environmental protection and the improvement of China's energy efficiency.[4] The Twelfth Plan also states that the government will establish a credit-rating system for enterprises' environmental behaviors and will step up credit support for those enterprises and projects that operate in line with environmental requirements and credit principles. The government will also build a green rating system in banks, linking green credit performance with other elements, including work evaluation of banking staff, admission of institutions, and business development. In 2015, however, the government had achieved most of Twelfth Plan goals both in energy and carbon intensity and in the share of renewable energy in the country's final energy consumption.

Because the measures under the Twelfth Plan were not delivering enough, the deepening urban air pollution crisis triggered a flurry of policy initiatives in 2013 (Jin, Yana, Henrik Andersson, and Shiqiu Zhan 2016, 4–5). From mid-June of that year, a series of regulations aimed at controlling air pollution was announced, the prosecution of environmental crimes was made easier, and local officials were told they were to be more accountable for air quality in their areas. At the same time, the government promised $275 billion of spending to clean up air pollution over the next five years, a sum that equates to twice the annual defense budget.

In October 2013, Beijing's city government unveiled a further round of measures—from restricting the number of cars on the road on any given day to closing schools—to be implemented only in the event of air-quality emergencies. A few days later, air pollution triggered by the start of the winter heating season brought Harbin, the capital of Heilongjiang Province, to a virtual standstill with closed schools, paralyzed public transport, and grounded planes. A new, tougher draft of environmental protection law came up for its third reading before the bimonthly session of the Standing Committee of the National People's Congress (NPC) in that same month, but the vote was postponed as legislators called for harsher actions to protect the already heavily polluted environment. The law had not been revised since its passage in 1989, despite China's deepening environmental crisis and the government's apparent inability to reverse the trend. The draft amendment reiterated the call for more responsibility and more spending on environmental protection at all levels of government and increased the weighting of environmental protection in governmental performance evaluation. Indeed, officials remain accountable of environmental degradation for a number of years even after the end of their mandate. It also included a contentious proposal to severely limit the civil society groups that were entitled to bring environmental lawsuits.

The new law will exert a great deal of pressure on local officials, who are evaluated by the central government. Until now (and even if the environmental performance criteria were in place), this policy had not really been enforced; local economic actors could easily buy the support of local officers to be exempted from compliance.

A REVISION OF INTERNATIONAL ENVIRONMENT DIPLOMACY

As China moves to a clearer environmental agenda and more stringent policies, will it change its traditional international position?

The environmental question has become a crucial and even unavoidable internal question for China, one to which the government can no longer give rhetorical or cosmetic answers. The environment's effect on the country's position in international negotiations is, however, not necessarily direct. China and its allies (including Brazil, India, and South Africa) still primarily defend a multilateral approach throughout all major negotiations, leading to an inflexible defense of the differential treatment of "catching-up" countries. Unlike the United States or the European Union, however, China's positioning on the international scene doesn't exactly follow the chronology of its internal evolution.

During its economic catch-up period, Beijing simply took a passive role in the international system and within the G77. It therefore adopted a classic third-world attitude, limiting its commitments, putting the burden onto historically developed countries, and requesting time extensions and technological transfers. The country eventually adopted a more active position under Hu Jintao's government (2003–2013); this position was enhanced once President Xi Jinping came to power in 2013, a time of increasing diplomatic aggressiveness for China. During this same period, the exit of the United States from international climate negotiations had already hampered multilateral processes and reaffirmed the primacy of national interests. China could thus adopt a similar position and defend—rather legitimately—its sovereignty on its own natural resources, as well as the protection of its economic development possibilities, without opposing the U.S. vision.

Chinese authorities constantly reaffirm that although the evolutions of their domestic policies do have global impacts (for example, from their carbon emission reduction policies and their massive reforestation programs), they are still only dictated by strictly national choices. China committed on a voluntary basis to reduce the carbon intensity of its economy by 2020 and invest in renewable energies (this was also a result of internal motivations). This position is consistent with the evolution of international negotiations, especially those on climate, where the "pledge and review" regime has prevailed over more centralized (top-down) models for the distribution of emission reduction efforts between countries.

However, China is in a transition phase. Even though it is still partially viewed as adopting the defensive diplomacy of developing countries within the G77, it has also taken a leading role in the group of emerging countries (BASIC—Brazil, South Africa, India, and China), which are increasingly active in the formulation of the rules of the game: "G-77 members have not helped to set up the rules for international institutions and events. The large regional powers among the BASIC countries, in contrast, have drawn increasing international attention and can play something of a veto role in global negotiations and rule-making, even if they cannot necessarily assert a positive agenda" (Hochstetler 2012, 53).

Indeed, the recent evolution of China on a number of global issues is striking. The agreement on HFCs with the United States is a significant sign of the different positioning of China during the Paris Conference on Climate Change in 2015. As a contribution to the Paris agreement, China accepted a reduction in its absolute level of emissions, with prospects of peaking CO_2 emissions in 2030 or before. This is a shift from the traditional position, as is the channel used before Paris to float these ideas. China's use of soft power on global environmental issues is entirely new. The control on foreign policy matters was traditionally very strict. In the past few years, however, the government has created institutes to deal with climate change, encouraged universities to engage in international cooperation in the field, and invited a number of international experts to participate in the discussion.

This new attitude is due to the country's new awareness of the risks, its willingness to contribute to a global agreement (which is a condition for China to reduce its exposure), and its desire to be recognized as a responsible citizen of the world community. This new stance has facilitated the engagement of China with the U.S.

government leading to an effective coordination between the two countries to prepare the final phase of climate negotiations and supporting each other until the final conclusion

LOOKING FOR A G2: THE U.S.-CHINA DIALOGUE

CLIMATE NEGOTIATIONS: REDUCING "BURDEN SHARING" TO THE EVALUATION OF EFFORTS, INSTEAD OF WORKING TOWARD A COMMON OBJECTIVE

For the United States, the measurement of GHG emission reductions has become a nonnegotiable condition for its participation in any climate agreement. Beyond taking stock of efforts, the United States promotes measurement reporting and verification (MRV). Indeed the capacity of checking each other delivery of action is a major element for trust building. The United States already undergoes rigorous reporting of its domestic mitigation actions in order to build the international community's confidence, and other countries, especially developing countries that receive financial assistance from the developed world on their mitigation actions, should do the same. This position was fought against by China, which, on principle, refused to submit its environmental efforts to international external expertise unless it also applies to actions financed by international support (Colombier et al. 2011).

The sixteenth session of the Conference of the Parties to the UNFCCC, held in Cancun in 2012, resulted in remarkable compromise in this area—notably, with the help of India. In order to be taken into account, efforts shall be verifiable—in particular, the efforts of developed countries, which will be appraised according to the GHG reduction targets announced in the Copenhagen agreement. This breakthrough seems be to be linked with the "method" promoted by Barack Obama: a nonstop effort to reach out to China to get a deal. First narrowing differences in Cancun through intensive bilateral negotiations between U.S. envoy Todd Stern and Chinese negotiator Xie Zenhua (*New York Times*, December 8, A 15), President Obama went on to note that the two delegations had met several times in the six intervening weeks between Tianjin and Cancun and that relations had been quite cordial. Similarly, the head of the Chinese delegation, Su Wei, dodged several opportunities to criticize the United States in a press interview, saying rather benignly: "We recognize that the U.S. is doing quite a lot after President Obama took office."

Unlike the abrupt positions of their predecessors, U.S. negotiators could thus take advantage of two major advances in their relationship with Congress:

- **The transparency of emerging economies' commitments**: with the Cancun decisions, all major emitters will have to publish regularly—at minimum every four years—their national communications and GHG national inventories. All countries should aim at doing it more frequently if they are able to do so. The Cop 16 Decision 1/CP.16 63. decides to conduct international consultations and analysis of biennial reports under "in a manner that is non-intrusive, non-punitive and respectful of national sovereignty; the international consultations and analysis will aim to increase transparency of mitigation actions and their effects, through analysis by technical experts in consultation with the Party concerned" (UNFCCC 1/CP16). Indeed, COP 16 agreed decisions continue blurring the strict differentiation between developed and developing countries following Copenhagen accord logic.
- **The parity of efforts**: A secondary goal was the legal parity in how the pledges of the United States and China were incorporated into the Cancun agreements. The United States achieved this goal by having both developed and developing country pledges "taken note of" in the final decision text (UNFCCC 1/CP16 and Morgan and Seligsohn 2010).

Better still, the United States and China embarked on an extensive dialogue on environmental issues from the standpoint of their respective national interests. This dialogue between the two major global GHG emitters covered the principal emission drivers on a reinforced cooperation for the harmonization of norms, as well as on new technological applications and developments (U.S.-China Climate Change Working Group 2014).

The main areas of concern are a reduction in emissions from vehicles; "smart grid" technology; carbon dioxide capture, utilization, and storage; and energy efficiency in buildings and industrial plants. More important, the two governments decided to share information regarding their respective post-2020 plans to limit GHG emissions and prepare their contributions to set a solid base for the agreement in 2015.

This bilateral preparation of a multilateral agreement was unprecedented. It was the result of intensive U.S. diplomatic efforts, which were increasingly welcomed by the Chinese government. This cooperation was de facto indispensable for the United States. Through an increased coordination with China (including under the forms of pledges for 2015), the U.S. administration aimed to obtain the most perfect parallelism between its own obligations and those of its major competitors—and hence respond to the Byrd-Hagel resolution.

China is concerned with its economic development and therefore with potential trade retaliations, which the government dreads. Since its most important trade partners—the United States and the EU—could consider its climate policies as restrictive and disadvantageous for their economies, the government could try to restore the nation's industrial competitiveness through trade measures as border tax adjustments, for instance. Thus, from China's perspective, the bilateral dialogue with the United States and the common elaboration of a balanced solution on the major intersections of environmental and economic issues is a security solution. The debate between the two global powers of this improbable G2 is thus entirely focused on the protection of national interests and the balance of power. China obviously has the determination to demonstrate its new strength and compare it to the American power.

G2 AS A GAME CHANGER

The balance of power undoubtedly underpinned the deal struck by China and the United States on November 12, 2014. China agreed to curb its emissions after 2030 and possibly before, while the United States announced a 27 percent cut by 2025 with reference to 2005 levels. This deal was negotiated under secrecy over a one-year period.

Although it did not come out of UNFCCC talks, it is very likely to be a game changer in climate change multilateral negotiations. At first sight, it postpones substantial emission reduction efforts beyond 2025 and 2030. Meeting the 2°C objective by 2050 would require an overall halving of global emissions, with a burden sharing of a roughly 80 percent cut for historical big emitters such as the EU and the United States. In the case of the United States, this would imply a 4–5 percent emissions cut per year between 2025 and 2050, as compared with a 0.45 percent reduction per year between 1990 and 2025, if the November 12, 2014, objective is met. Yet, when it is broken down into two periods, this average figure masks a clear shift downward in the emissions trend. The trend in emission reduction would be much steeper in 2005–2025, thanks to the 2025 objective, than during the 1990–2005 "lost period," meaning that the United States is moving toward decarbonization. The same holds for China. For the first time ever, China announced to the world—and not just the United States—an emission peak. China is now accountable.

In September 2015, the United States and China repeated the bilateral strategy with the publication of a new statement. This one intended to define comparability in transparency of climate action in both countries, a progress compared with the Copenhagen/Cancun agreement. This was a win for the Obama administration and a long-sought response to the Byrd-Hagel resolution. Through intensive bilateral negotiation, the United States finally found an agreement with China, overcoming a major obstacle for a global deal. The U.S.-China joint announcement on climate change in September 2015 states the need for "the inclusion in the Paris outcome of an enhanced transparency system to build mutual trust and confidence and promote effective implementation including through reporting and review of action and support in an appropriate manner" (White House Office of the Press Secretary 2015).

The two giants had moved forward, experimenting on environmental issues and designing what a G2 could look like. Surprisingly, their experiment did not result in an undermining of global framework. Indeed, it defined boundaries within which the two countries could agree to negotiate. A staged approach was also adopted by several negotiating groups on the eve of the Paris agreement, allowing identification beforehand of the win sets of major players. Thus, the G2 organized around climate change policy may have generated a new form of multilateralism.

The 2016 U.S. elections put an end to that experiment. The Trump administration, supported by a Republican-dominated Congress, has taken aggressive action against many environmental regulations—even attacking the Clean Air Act, the bedrock of U.S. environmental regulation and deciding to withdraw from the Paris agreement. Even if the intention of deregulation is clear, however, it is far too early to assess the depth and effectiveness of such a move. Different checks and balances can temper such an aggressive stance: State policies, such as in Texas and California, will diverge on the support to renewable energy, energy efficiency, and carbon prices. In addition, many actions in court will delay or hamper drastic moves on environmental deregulation. Some global businesses will oppose what they perceive as acts that undermine their competitiveness. The scientific community and social movements will also attempt to mobilize public opinion. Nevertheless the domestic driver of U.S.-China partnership is now blocked.

On China's side, there is no sign of decreased commitment to the environmental agenda. President Xi Jinping forcefully expressed China's intention to implement the Paris agreement and called for a responsible attitude from the United States. Several decisions on coal consumption led to the canceling of more than a hundred coal-fired projects—amounting to a 120-GW capacity. Closing coal mines makes credible a peak in emissions much earlier than announced in Paris. Still, China cannot fill the void left by the United States, and it needs allies: Thus, it is Europe that must now stand up and work to keep China on the move.

12

THE NEW MULTIPOLARITY OF SUSTAINABLE DEVELOPMENT

FOLLOWING THE analysis of U.S. and Chinese global politics in chapter 11, this chapter describes other dynamics at work in the global environment domain. We draw our analysis from the climate change regime, as this is the most active and politicized field in which countries try most to exert their influence.

One trend is particularly significant: the crisis in the climate talks at Copenhagen revealed the new status of emerging powers and the end of uncontested U.S. leadership. The old world was declared over through this amazing political sequence in which coordination between Brazil, South Africa, India, and China (BASIC) defeated U.S. and European proposals. This new multipolarity of the international system seemed to appear as a consequence of the catching-up of some countries—in particular, the four members of the BASIC group, which mainly united in opposition. But the reality is more complex. Since Copenhagen, the climate regime has witnessed a fragmentation of the group of developing countries, with new coalitions organized around more progressive positions that are committed to sustainable development and strong climate action.

The European Union has played an important role in helping these coalitions emerge. In doing so, it has evolved from a position of front-runner, leading by example, to facilitator of a broader movement. This chapter focuses on the reassessment of the EU's role in the global scene and the fragmentation of the developing countries group. It looks at the emergence of progressive countries in favor of green development, which are challenging the classical positioning of G77 and China that had been established in January 1985. It helps explain why, despite many economic difficulties, the trend in favor of sustainable development is not only maintained but is progressively mainstreamed in the domestic policies of many developing countries.

The geopolitics of the environmental field can be analyzed as a case study of power politics and interest-based strategies that are not much different from what happens in the security or trade domains. The confrontation and cooperation of the two major economies—the United States and China—within the climate regime can be explained largely through their interests in preventing one another from free-riding a potential climate agreement that they both need. Even in this case, however, power politics and interests fall short of explaining the entire picture: the change in the attitude of Chinese leadership, from neglect or reluctance to active participation in the climate talks, comes from a profound revision of its development doctrine.

The role of European politics and climate change policies is even clearer when one analyzes how certain key ideas shaped negotiating positions. The EU's early positioning on active climate action was due to a specific political context in which green parties participated in many European governments in the 1990s. It was also due to the influence of the more advanced policies of Nordic countries. In the early 1990s, the newly created EU positioned itself as the new leader in the global environmental domain, in a shift from previous U.S. dominance. The EU had even produced a theory of this leadership: "leading by example." It built on its own progressive constructions of rules and norms, which are the center of gravity of European governance. Governing by norm setting is thus the foundation of the EU's behavior both internally and externally. It therefore decided many times to act unilaterally for the common good, whereas other big powers chose to carefully determine their actions in comparison with others.

Europe is at a turning point for both internal and external reasons. Internally, there is an increased critique of such unilateral policies because of the costs inflicted on the European economy when other competitors don't act similarly. Externally, there are rising doubts about the EU's capacity to pursue this path and a dispute by some of its leadership. Now is the time for the EU to change its behavior, to look for allies and coalitions to maintain the value of its contribution. A major opportunity is offered by the new and very different framing of collective action on climate. The preparation of the Paris agreement and its conclusion shifted the discussion from carbon emission reduction targets to a broader discussion on energy transition and economic transformation. In Europe, this shift was necessary for internal reasons to prepare the post-2020 climate policy and to overcome the divergences between countries that rely on very different energy sources. However, the broadening of the perspective also allowed the broadening of partnerships. Europe could now speak in terms acceptable to all countries, including developing countries, as every country could try to respond to the challenge of climate in its own way.

A successful global action for the environment will effectively depend on a sufficient number of countries moving forward, introducing significant changes in production and consumption models, and not waiting for others to act. The European perspective could then be a shared ownership by a broader constituency of like-minded countries.

THE EUROPEAN UNION: FROM LEADERSHIP TO COMPANIONSHIP

Environmental protection has been an essential element of European integration for more than thirty years, as expressed by the founding treaties of the European Community and later of the European Union. The Single European Act in 1986 established the environment as one of the three competencies of the European Economic Community (EEC). The Maastricht Treaty of 1992 made environmental protection one of the EEC's objectives, alongside the completion of the internal market and the improvement of European cohesion. Finally, in 1997, the Treaty of Amsterdam added a reference to sustainable development among the EU's objectives, which later inspired the Lisbon Strategy of 2000 and the Gothenburg European Council's decision in 2001 to support the Kyoto Protocol, despite U.S. withdrawal.

Because Europe has been building supranational environmental policies from 1986 to the present, it has witnessed the benefits of such transfers of sovereignty, since national environmental policies in Europe are highly dependent on transboundary agreements. For example, improvements were noted in acid rain and hence in the water quality of the Rhine and the North and Baltic Seas. It was also easy for Europe to actively participate in the global discussion when three global conventions for the protection of the ozone layer, the climate, and biodiversity were agreed upon (Lönnroth 2006). Moreover, the defense of a multilateral system of norm setting was the natural way for the EU to defend the European model, having championed at home a system of collective action based on norms and regulations in exchange for its members having partly given up their sovereignty.

The European project embodies the aspiration that states can and must work together for the common good more than ever before in history (Grubb and Gupta 2000). Action to support the common good beyond national interests is reflected by the norms that have become conditions for accession to the EU and allocation of Community funds. The European commitment to global environmental goals is founded on concrete applications under the structure and use patterns of European budget lines.

A MUCH-NEEDED LEADERSHIP IN CLIMATE ACTION

Even though one can find such a trend throughout the whole realm of environmental negotiations, it is undoubtedly in the field of climate change that the EU has been most active in trying to exert leadership and impose an international regime.[1] In 1996, even before adoption of the Kyoto Protocol, the EU Council of Ministers put 2°C forward as the maximum limit for a rise in temperatures. Other achievements over the years include the following (Tubiana and Wemaëre 2009):

- It has continued to support ambitious targets for the climate regime.

- It developed an integrated, though comprehensive, set of policies and measures identified through the European Climate Change Program (ECCP), which was launched by the Commission in March 2000.
- It showed concern about the environmental integrity of the Kyoto Protocol's flexible mechanisms (in addition to international offset credits for complying with emission reduction targets and reluctance to use forestry credits because of the temporary and reversible nature of sequestration through forestry activities).
- It developed the first mandatory regional emissions trading scheme in 2005, with the launch of the EU Emissions Trading System (ETS), which puts a price on carbon for key EU emitting sectors (i.e., power generation and energy-intensive industries).

The establishment of the EU's leadership also benefited from the American retreat from climate negotiations under George W. Bush's presidency. The EU has gained significantly in stature and international recognition, especially since 2001, thanks to the rescue of the Kyoto Protocol following the 2001 U.S. exit and the 2004 EU-Russia deal on Kyoto (Afionis 2010).

The European position on climate change has thus been considered emblematic of its overall international position: "Climate change is not only an extremely grave international problem but an issue which had been acquiring a symbolic profile in the protest movement against the destructive effects of globalization. It may therefore be affirmed that this agreement on the Kyoto Protocol was of major importance in terms of the regulation of globalization and international governance" (European Parliament 2002, 7–8). This was stated even more clearly by Margot Wallström (2001), European Commissioner for the Environment, after the Gothenburg European Council in June 2001: "I think something has changed today in the balance of power between the U.S. and the EU."

In truth, when the EU proclaimed itself leader of the fight against climate change, it is also attributable to the fact that, at the time, nobody else really wanted to fill this position (Martin 2012). The EU has the merit of having kept the process of negotiating a climate regime alive, even though the latter was viewed by many as already moribund (Victor 2001). However, the EU's momentum on environmental issues in the early 2000s slowed after the entry into force of the Kyoto Protocol, thanks to a changing global context and shifting internal balances within the Union.

TIMES OF CRISIS

The EU's capacity to retain a leading role has been challenged for two main reasons: the effects of its enlargement on the internal consensus on environmental action, and the economic crisis that has unfolded since 2008.

The 2004 enlargement resulted in the integration of countries whose primary concerns were not environmental issues, because of their level of development and their economic structures. After this enlargement, one can observe a severe loss of interest in environmental questions, not only within the Commission but also among governments of several member-states. Eastern and Central European countries had difficulties adopting the environmental chapter of the *acquis communautaire*, the common body of legislation. The widespread use of transition periods for adopting the *acquis* represented a real risk of a partial "renationalization" of environmental policy making. Avoiding such a renationalization required enormous investments of time and resources, which would ensure the compliance of new member-states to the existing European legislation.

Even though the adoption of environmental policies *in fine* had a positive influence on the new members' environmental behaviors and strategies, which would otherwise have been engaged on less favorable paths, it required a considerable effort that had important political and economic costs.

EU IS NO MORE ALONE AND BEGINS TO STRUGGLE

These difficulties were particularly striking during the attempt to revive environmental policies at a time when the European Commission wanted to launch preparations for the second phase of the Kyoto Protocol (the first period of which was expiring in 2012) and for the international negotiations of a more global agreement in 2009.

To play a leading role in these negotiations, the EU wanted to develop a common position in the fight against climate change; it thus implemented its own measures to deal with the issue. The European Council of December 2008 definitively adopted the package, named the "20-20-20 targets": by 2020, reduce by 20 percent the emissions of greenhouse gases, increase by 20 percent the EU's energy efficiency, and reach 20 percent of renewable energies in the EU's total energy consumption. In addition, the emissions allowances of greenhouse gases had to be fully auctioned, though within a gradual schedule: companies would have to buy 20 percent of allowances from 2013, 70 percent in 2020, and 100 percent in 2027. In the electricity sector, exceptions would be accepted for new member-states until 2020, while the auctioning of all allowances for electricity would be effective from 2013 for other EU members.

For the first time, however, the EU also introduced a conditional clause for the implementation of its policy: emissions reductions could be extended to 30 percent (compared with 1990 levels) if a sufficient international agreement could be reached in Copenhagen. The industrial companies most exposed to international competition would otherwise benefit from a free allocation of quotas to avoid the outsourcing of the most polluting industries, subject to competition of arrivals from countries with little involvement in the fight against global warming.

Poland and most new member-states, where electricity relies mainly on coal, feared that this reform, which increased electricity prices, could undermine their economic growth and their energy security. Poland and the Baltic States, in particular, feared that the package would force them to develop their gas imports from Russia to reduce their GHG emissions, thus limiting their energy independence. The collective response to these concerns was the decision to improve electrical interconnections within the European electricity market. The European package was approved, but the difficulty in negotiating it resulted in a core group of countries reticent about the pursuit of an active climate policy.

Successive economic crises finally affected not only the EU's capacity to respond as a supranational entity, but also the capacity of states to tackle the deficits caused by speculative bubbles (in the banking, financial, and real estate sectors, etc.) and to reduce public deficits through more stringent budgetary policies. Concerns about strategic energy security began to dominate—sometimes in contradiction to emission reduction objectives. While switching from coal to natural gas is a typical example of abatement, the risk of depending on imported gas in a difficult political context secured coal's role as a pillar of electricity production in the EU. The same applies to the problem of carbon leakage—the loss of European industries' competitive advantage over their foreign competitors has become an inescapable part of every climate policy. Thus, the EU policy cannot be more stringent than its competitors, unless it protects itself through the adoption of trade measures, such as border adjustment measures or exemptions for its more exposed sectors.

Even if the fears of competitiveness losses were overemphasized, they induced a restraint in the full implementation of climate policy instruments. The most affected instrument was certainly the EU ETS (the regional carbon market), as demands for overallocation of emission rights and the slowdown of the European economy resulted in a collapse in carbon prices. Today, carbon prices remain well below any meaningful signal, and a deep reform of the system is needed. On the national level, as well as on the EU level, environmental issues have been sidelined, even though major decisions on environmental matters were not reconsidered, such as the reforms of the Common Agricultural Policy or major common research programs.

Still, the EU has played a major role in the aftermath of Kyoto Protocol, which has been abandoned by several countries. Indeed, it has maintained pressure in establishing regulations, standards (e.g., for car emissions), and emissions-trading schemes, thus promoting renewable energy and energy efficiency. The main features of EU climate policy, even in the international cooperation domain, have spread throughout a growing number of countries, including China. For example, emissions-trading schemes have been adopted in various countries (New Zealand, Switzerland, and soon in China) and states and regions (California, Quebec, the northeastern United States). In addition, many more countries, regions, and cities are considering launching emissions trading or putting a tax on carbon (for example, the Carbon Pricing Leadership Coalition, launched by the World Bank in 2014). Over the past ten years, car fuel standards, targets in renewable energy in the energy mix, green procurement policies, and so forth can be found in countries around the world (Bals et al. 2013).

The Cancun pledges, in which ninety countries took commitments of emission reductions by 2020, and the 189 Intended Nationally Determined Contributions (INDCs), presented before the Paris agreement in December 2015, showcase the diffusion of ideas, policy instruments, framing of targets, and objectives that the EU developed over the past fifteen years. The EU is no longer alone in its climate change work. Thus, the EU losing part of its leadership actually comes from the pioneering strategy of "leading by example," which worked very efficiently, even if the cumulative effect of its actions is still inadequate.

It is now more difficult for the EU to rely mainly on its internal dynamics for its reestablishment as an international leader. Nevertheless, the negotiation of a post-2020 climate regime represents an opportunity to redefine both its domestic policies and its international strategy.

POST-2020: FROM CARBON TARGETS TO ENERGY TRANSITIONS

The European Union—just as any other nation—had to prepare its proposals for the Conference of the Parties in 2015, at the end of which a new global agreement was negotiated. The conference took place within an unprecedented setting, as the obligations of the Kyoto Protocol were to be replaced with commitments by all countries. The EU no longer had to define itself in relation to developed countries only, but rather within a wider range of countries: it thus had to propose a structure and rules for a universal agreement. The COP in 2015 was a double exercise for the EU: it consisted of developing a European policy for the period after 2020 and of advocating for a global framework that would meet its vision and interests.

In 2020, Europe should have reached, by a wide margin, the objectives it adopted in 2008: carbon emissions will be reduced by more than 20 percent compared with 1990 levels, and the share of renewable energies in the energy mix will be superior or equal to 20 percent in many countries. Only the nonbinding objective of improving energy efficiency will probably be hard to fulfill. This climate and energy package was criticized, particularly in the context of the economic crisis suffered by Europe since 2008. European industry and consumers questioned the energy policy, which has been accused of raising energy costs and jeopardizing firms' competitiveness through the price of carbon and financial support to renewable energies. The successful exploitation of shale gas and oil in the United States (though it glosses over any environmental impacts) and the following decrease of natural gas prices have reinforced this criticism.

The European energy policy therefore must adapt to a new context in which the fight against climate change is presented as an obstacle to growth. In this sense, the European debate and negotiation reflect the state of international discussions. The reconciliation of emission reduction targets with economic growth remains the condition for a global agreement and its implementation.

Net imports of fossil fuels do weigh heavily in the European balance of trade, representing about 3.3 percent of the twenty-eight member-states' GDP in 2013. Europe had to build a proposal for a post-2020 policy that would meet the energy security imperative, which can be ensured by a more coordinated procurement policy and a special effort to improve energy efficiency—still one of the best means to reduce dependence on imports. The negotiation of the new European climate and energy package has mainly focused on an objective of "at least" a 40 percent reduction in GHG emissions, compared with 1990 levels, by 2030. The objective of a 27 percent share of renewable energies in the energy mix by 2030 (with a possible upward revision to 30 percent) is also a part of the final package, as are energy-efficiency plans. Since the economic crisis, European negotiations are working like international climate negotiations: member-states want to ensure the autonomy of their energy choices and avoid excessive costs.

The assertion of national choices, which can be divergent with regard to renewable energies, resulted in a different process of building a European position. The EU's position became more bottom-up, as it included elements of support for emitting or fossil fuel–dependent countries in the policy packages. Its purpose is to make possible national energy transitions in different countries that have various resources and economic structures. The paradox, however, is that the major emitting countries in Europe, where emission reductions would a priori cost the least, since important efficiency gains are possible, are the most reluctant member-states to make commitments. The carbon lock-in—notably, in relation to energy infrastructures and electricity production—and the necessary organization of job retraining programs are major bottlenecks.

To trigger such structural transformations, the decarbonization of the economy must rely on energy policies that cover all sectors: from energy production to energy demand, from energy consumption patterns to the composition of the energy mix itself. Most European countries, including Germany, France, Denmark, and the United Kingdom, have engaged in the development of comprehensive energy transition policies. The more far-reaching decarbonization objectives are, the more important the required structural and technological transformations, and the wider the range of policies that must be led. In the case of Europe, decarbonization targets that are consistent with the international community's objectives are very ambitious: at least 80 to 90 percent emission reduction by 2050 compared with 1990 levels. Some front-runners (countries like Sweden and France) have already announced that they will reach an objective of net zero carbon emissions around or before 2050. It is therefore only logical that climate policies in Europe are not confined only to price instruments (taxes and carbon markets) anymore, though they remain crucial; they also encompass technological, financial, and regulation aspects, especially with regard to energy markets.

FROM LEADERSHIP TO COMPANIONSHIP

The EU's enlargement also made the difficulties of European institutions within international negotiations even more visible. Once enlarged to twenty-five member-states, the EU was neither a state nor an orthodox international organization. During climate negotiations, it was routinely treated as an actor capable of autonomy and volition; at the same time, however, the differences between its member-states were manifest, and coordination processes were extremely costly. The EU's internal operational system sometimes makes it difficult to reach a consensus, especially with the rotating presidency. The fact that the sitting president needs to keep contact with the previous one and the incoming one to coordinate efforts and agree on common discourses can be "controversial" (Martin 2012, 200).

EU policy is indeed the product of compromises among its now twenty-eight states, which are characterized by different income levels, economic structures, and energy mixes. While the transition to a low-carbon economy and the cost of climate change after 2050 is on the European Commission's agenda, member-states remain concerned about the short-term preservation of existing businesses and jobs for voters (Tubiana and Wemaëre 2009). The Copenhagen Summit was an illustration of these trends. There were divergent interests in member-states and conflicting positions about the role the EU should play, including the financial resources, the type of commitments that could be signed, and where the focus should be put—for example, climate change (France and Sweden) or energy security (Czech Republic). In the end, the common position was agreed to at the last minute, with only a minimum common denominator. Because it was so last minute, the EU was not flexible enough to play a role in the final deal, as every movement needs to be consulted with all member-states. This eventually prevented its ability to react in time. Meanwhile, the United States and major emerging countries defined the compromise they wanted, dropping major concerns of the European position. In Copenhagen, the EU was only reacting and not proposing (Martin 2012).

The lack of leadership at Copenhagen has been mitigated by the fact that the EU succeeded in finding common positions and establishing internal targets on emission reductions. In this way, the EU moved toward a low-carbon economy and has continued to play an exemplary role. The coordination of member-state policies remains strong, and its efforts at various scales of the EU's territorial governance are still important, which allows it to uphold not only its status but also common international positions.

Yet, the Copenhagen final agreement was viewed as disappointing—or even a failure—by European leaders; thus, the EU now finds itself, just like other nations, lowering its ambitions in order to achieve success. According to Lisanne Groen, Arne Niemann, and Sebastian Oberthür (2012), who compared the EU's role before and after Copenhagen: "In contrast, at the 2010 Cancun negotiations, the EU arrived with much more limited goals and advocated a set of concrete decisions to implement various elements of the Copenhagen Accord. The EU also established itself in a bridge-building role between different blocs of countries, actively building coalitions based on its position on each issue."

Since then, this course and approach became even clearer. After Copenhagen, the EU has been actively looking for allies—notably, from small islands, Latin America, and Africa—who might share its ambitious emission reduction targets, in line with the 2°C global objective, as well as its strong, transparent rules of implementation. During the Durban Conference in 2012, the EU built on its new allies to revive the process of negotiating a global agreement succeeding the Kyoto Protocol. At COP 21 in Paris, this strategy of coalition building went well, allowing the EU together with small island states to launch the "High Ambition Coalition"—a group of states that played a decisive role in the finalization of the agreement.

Indeed, Europe has learned from past errors, and the two-stage process—first, internal coordination and after, negotiation with others—does not prevent Europe from exploring potential landing zones with others. Europe is now more flexible and better equipped to explore, through informal processes, vision and compromises that better integrate demands from "others." The EU's pre-Paris activities were indeed intense. Informal discussions with China took place on various topics—in particular, transparency and possibilities to periodically revise national commitments to bridge the gap between emissions trends and the goal of limiting temperature rise below 2°C. More important, Europe developed a strategy with Latin American countries that resulted in new coalitions: first at the Cartagena Dialogue and later on the eve of Paris talks the "high ambition coalition" of thirty-five rich, poor, and emerging economies.

THE FRAGMENTATION OF DEVELOPING COUNTRIES AND THE CHALLENGE OF MIDDLE-INCOME COUNTRIES

The international political landscape has become increasingly fragmented since the end of the Cold War. In the environmental field, this fragmentation is most apparent. The traditional clustering of the G77 and China, established in June 1964 during the first United Nations Conference on Trade and Development (UNCTAD), which currently includes 133 countries, is not the only regrouping federating the interests of developing countries. More than fifteen negotiating groups currently exist, and new ones are regularly created.[2] This multipolarity reflects the weakening of major countries, as well as the incapacity of one single country to play a hegemonic role (Roberts 2011). It is also—and above all—the expression of increasing divergences between the interests and concerns of developing countries. This diversity of interests has resulted in a proliferation of groups focusing on one specific aspect of the climate negotiations. The case of insular regions is a good example of this evolution: the constituted coalitions, such as the Alliance of Small Island States (AOSIS) and the Small Island Developing States (SIDS), are handling the impact of climate change on these territories and are organized as lobby groups in an attempt to obtain more stringent mitigation commitments and a more ambitious global goal (1.5°C against 2°C as the stabilization goal of global warming). A great number of countries are participating in several such groups, all of which are attached to particular issues (e.g., landlocked countries, forests, vulnerable countries). They reflect the complexity of a climate regime in which the negotiations and actions cover emission reductions efforts, as well as impacts of climate change, industrial and energy policies, and forest and agriculture.

EMERGING MIDDLE-INCOME COUNTRY ACTORS: PROACTIVE OR DEFENSIVE

The fragmentation of the G77 can be seen in particular in the middle-income countries. The most visible group is the one gathering the biggest emerging countries, constituted as the BASIC group (sometimes also associated with Russia), which coordinates its positions on a regular basis. The BASIC group was organized as a response to developed countries' demands to limit their emissions and to defend their concerns and priorities: economic catch-up and preservation of national sovereignty. At that time (in 2009), these countries considered environmental, and in particular climate, negotiations a zero-sum game. In fact, these countries are under pressure not only from the most developed countries but also from the rest of the developing countries, since the constant growth of their emissions makes the objective of stabilizing the climate unattainable. The success of a climate agreement in the long run depends mostly on the absolute reduction of their emissions, which seems highly difficult to achieve, unless they engage in a profound transformation of their energy systems and development models.

This group is thus in a very defensive position, attempting to limit constraints on its development and advocating for a binding framework for developed countries that is simultaneously flexible for them. Along with the United States, this group is a powerful player, able to define a negotiation's point of equilibrium. It is a new element of international negotiations, and a crucial one for climate negotiations, where, even more than in other fields, strategies must be tightly coordinated and interests carefully aligned.

This group has nevertheless experienced growing divergences: China has developed its own approach, which is often not aligned with its partners. Brazil, a low emitter in the group, has taken a more proactive approach in identifying a low-carbon economy as a positive factor for its economy, as well as participating in the High Ambition Coalition, a group that bridges different economic interests. India, a low emitter per capita, is struggling with increased energy demand, largely relying on coal-based electricity, and it is thus careful to find a balance between coal and renewable energy sources.

MIDDLE-INCOME COUNTRIES: PROACTIVE STRATEGIES

The other middle-income countries, which are not part of this "defensive" regrouping, are seeking to play a different role. Most of them are

Latin American and Asian countries, which have experienced important growth rates during the past two decades and which now wish to have international influence. They aspire to expand this influence by playing a "broker" role between developed and developing countries, while using their domestic policies as a showcase for their international policy. They play on a different note than the BASIC group and rather appear to be challengers: for them, the environment is not a zero-sum game, and they expect to gain from international collective action. Coordination brings with it a global benefit, and the sharing and development of knowledge reconfigures interests in the long term. Furthermore, because they emit fewer greenhouse gases, they are also less subject to pressures.

Since the Copenhagen Conference, some of these countries have defended the idea of a transformation of economic models toward a low carbon economy. By developing a new thematic, offering a radical departure from traditional north-south divisions, these countries are attempting to trigger international action by showing the way, rather than by a discourse invoking the historical responsibility of the north, which used to make prior action by developed countries a condition to all commitments by developing countries.

BROKERS NEEDED BETWEEN NORTH AND SOUTH

Several of these countries are indeed brokers that seek to build a consensus on the basis of new ideas and concepts. Mexico successfully played this role during the Copenhagen Conference, by launching—along with Norway—a campaign for the establishment of a green fund that could support and finance the transition of developing countries toward a low-carbon economy. This fund is original in its conception: initiated by two countries with different development levels, it broke the traditional codes of international development financing, which was usually focused on official development assistance led by donor countries. It thus postulates that the north and south are equal partners of international governance. Moreover, it included international sources of funding (including revenues from the auctioning of emissions rights and other international sources), national funding, and private funding in the design of projects. The

Cancun Conference of the Parties in 2010 finally decided to establish the fund, the instruments of which were negotiated and finalized in 2014. Mexico achieved a key diplomatic success by situating itself in an entirely new area. President Felipe Calderón's personal investment in this project demonstrated that the field of climate change is considered a strategic domain of international relations where influence can be gained.

Another example of an active brokerage activity is the constitution of the Cartagena Group. This group for a progressive dialogue on climate change was created on the margins of the Copenhagen Conference by developing countries (Colombia and Costa Rica) and developed countries (Australia and the United Kingdom). The aim was to develop the potential of small to middle-sized powers from across the development spectrum to work together to overcome blockage in the process. The idea was that a group of countries could draw on the voices of compromise in the negotiations and empower the middle ground. This gave birth to the Cartagena Group, launched in 2010 in Colombia and now regrouping forty countries sharing an objective to "block the blockers." The group resists polarized positions and works to find pragmatic solutions. It includes members from Latin America, Africa, Asia, SIDS, Europe, Australia, and New Zealand, with a strong Latin American presence from the outset, including active participation from Colombia, Costa Rica, Chile, Peru, Guatemala, and Panama.

The Cartagena Dialogue advocates shared responsibility, promoting an approach that would see all countries reduce their carbon pollution commensurate with their relative economic capacities. The impact of such a coalition has been positive on all the conference of the parties since then—especially in Cancun and Durban, where a process to negotiate a global and legally binding agreement was launched.

DOMESTIC POLICIES AS AN INTERNATIONAL ASSET

Middle-income countries, which have new investment margins for their development, decided to promote sustainable growth and to implement innovative national policies. This decision met their aspirations to have a greater voice on the international scene. They also consider this reorientation of their development models as an investment for a better valorization of their natural resources and a more promising integration into the world economy. Economic openness, which most of these countries decided to adopt twenty years ago, doesn't seem able to ensure a real economic catch-up. Instead, the choice for sustainable development, green economy, and high innovation seems to be the best pathway to escape the middle-income trap.

Colombia financed a rapid bus system, which in 2000 in Bogota introduced an alternative to cars in its rapidly expanding cities. In 2008, Chile was the first Latin American country to set a renewable electricity production target of 20 percent of energy provided by electric power facilities by 2024. Costa Rica invested more than $400 million in environmental service programs—notably, in the field of forest protection—and has committed to be carbon neutral by 2050. In 2008, at the Fourteenth Conference of the Parties, Peru became the first developing country to offer a voluntary emissions reduction target. Peru's pledge includes reducing the net rate of deforestation of primary forests to zero by 2021. In June 2012, Calderón signed Mexico's General Law on Climate Change, which includes targets to reduce GHG emissions 30 percent by 2020 and 50 percent by 2050, while acquiring 35 percent of Mexico's energy from renewable sources by 2024.

As another example, South Korea, already a member of OECD, engaged in a "green growth" policy, which it used as an element of its international promotion. Korea indeed put several of its initiatives forward as its renewable energy policy, including biomass mobilization, recycling policies, the development of a clean car program, and a vast energy-efficiency plan. Building on these initiatives and an intense campaign on its green growth policy, Korea succeeded in being given positions in new international environmental institutions. It eventually obtained the headquarters of the Green Climate Fund (GCF) (Manzanares 2017). South Korea backed, on a voluntary basis, the implementation of emission reduction actions in developing countries and contributed to proposals on a registry of actions and MRV procedures that should logically take part in the pledge-and-review system, which characterizes the current climate regime.

The change of government in Korea, with President Park coming to power, did not prompt an official reconsideration of this orientation. On the contrary, the current government—as well as environmental organizations—has criticized the policies of the former regime, which was accused of being not protective enough of the environment. Critics essentially focused on the gap between the exterior image and the reality of the domestic policy, but not on the value and opportunity for Korea to defend sustainable development (Seong-kyu 2013). Newly elected president Moon Jae announced that he will accelerate South Korea's transition to clean energy (*Korean Herald*, May 10, 2017).

THE IMPACT ON INTERNATIONAL PROCESSES: A NEW COALITION

Middle-income countries do not content themselves merely with being brokers. Since the Doha Conference in 2012, some of them have created new negotiation groups. Six countries encouraged by the results of the Cartagena Dialogue—Colombia, Costa Rica, Chile, Peru, Guatemala, and Panama—with the support of the Dominican Republic, announced the creation of a new negotiating bloc: the Association of Independent Latin American and Caribbean states (AILAC).

These countries represent a revolt of the middle, neither the poorest nor the richest. It is rather a group of developing countries willing to step out from the shadow of the G77 and China and offer real ambition: The alliance is formed to counter the growing influence of BASIC and ALBA groups of countries, which does not reflect the economic realities of all the countries. The group has garnered support from some major parties such as Mexico and the civil society, who see it as a constructive way forward.

This group of countries wants to play its part. According to the Inter-American Development Bank (2012), Latin America's GHG emissions for 2012 were estimated to be 11 percent of total global emissions. On a per-capita basis and in proportion to the size of its economies, Latin

America contributes more GHG emissions than China and India. AILAC countries decided to stop waiting for emission reductions or financial support from wealthy countries and made the case for low carbon development at home and abroad. As Mónica Araya, a Costa Rican negotiator, explained in an interview in the Spanish newspaper *El Pais*: "[The negotiations are] always told as a battle of North versus South, rich against poor, but each time this explains less and less of what's happening. . . . There is an alliance of countries that want all nations to take on binding obligations, and that the negotiations process is adapting to a changing world" (Edwards and Timmons 2012).

AILAC's strategy reinforces the visibility of these middle-size countries on the international scene and nourishes the dialogue between the EU and Latin American countries on new bases, which have built the ground of a common strategy during climate negotiations. Indeed, when AILAC launched a new coalition—the "high ambition coalition"—in December 2015 to demand an ambitious agreement consistent with scientific studies, the EU, United States, and Brazil all joined immediately, distancing themselves from the BASIC group. Namely, this coalition has a specific purpose: to include a reference to a temperature limit of 1.5°C as a goal and a net zero emission global goal by the second half of the century. This coalition made the fragmentation of G77 and China a political reality, while new alliances were born with promising future developments.

SUSTAINABLE DEVELOPMENT GOALS: A NEW LEADERSHIP FOR THE MIDDLE COUNTRIES

Rio+20 is another illustration of an international cooperation process marked by the influence of middle-income countries. The anniversary conference of the 1992 Earth Summit was suggested by Brazil's President Luiz Inácio Lula da Silva during the sixty-fourth session of the UN General Assembly in autumn 2009. Its organization associated representatives of both north and south—a Frenchman (Brice Lalonde) and the former minister for the environment of Barbados (Henrietta Elizabeth Thompson). The preparatory committee was presided by Korea and Antigua.

One of the conference's topics was the condition for a transition toward a green economy in the context of sustainable development and the eradication of poverty. This decision rekindled historical conflicts, as southern countries felt suspicious about new principles, which were understood as new constraints to their development or even as an attempt to commodify nature. The negotiation therefore had little chance to succeed, to escape the traditional bottlenecks of multilateral procedures, and to avoid resulting in the minimalist approach of international agreements so often defended by major emerging countries. As the latter didn't want to take on external constraints (which would add to their domestic challenges), their influence could be tracked in the Rio+20 declaration through the omnipresent reference to the principle of "common but differentiated responsibilities" and the reaffirmation of the primacy of national sovereignty.

It was eventually the middle-income countries that saved the conference and provided a solution. If an agreement was finally concluded, it is thanks to the mediation of countries such as Colombia and Guatemala. The main step forward made at Rio+20 was the decision to prepare the implementation of Sustainable Development Goals (SDGs) of universal character after 2015—that is, not limited to developing countries. It is entirely ascribable to them. The idea for the SDGs was originally proposed in July 2011 by the governments of Colombia and Guatemala at an informal government meeting in Solo, Indonesia. They then gained considerable political momentum. The two countries led a successful diplomatic offensive during the General Assembly of the UN in September 2011 and built a coalition of countries supporting their cause before Rio+20.

These two Latin American governments see this proposal as the first step in the establishment of common guidelines for the future of national sustainability policies. The objective is to build a goal-oriented framework that will help guarantee the implementation of Agenda 21 and the Johannesburg Plan of Implementation and that is the integration of the environmental agenda with economic development. The SDGs contribute to positioning the three transversal areas of environmental, economic, and sociopolitical sustainability in all the activities of the UN system.

Colombia, also on behalf of Guatemala, emphasized that the SDG architecture should be tailored to reflect the multidimensional challenges faced by countries. Colombia supported a voluntary, flexible, and dynamic "dashboard" of targets and indicators, which countries could sign up to in accordance with national goals and priorities. According to Alejandra Torres, director of international affairs at the Ministry of Environment and Sustainable Development of Colombia and one of the leading negotiators: "SDGs had to be universal—the goals should apply to all, not just to some. Thus the focus of the MDGs on poverty reduction would remain central to SDGs, but would be complemented by targets that could incentivize action on sustainable consumption and production" (Watt 2012, 1). Colombia and Guatemala fulfilled their objectives: the negotiation on global SDGs was launched and led, after one year of work, to the presentation of a first document by the two cochairs of the Open Working Group on Sustainable Development Goals.

Do all these actions lead to a convergence of sustainability policies? At least the existence of these advocates of global action for sustainability is a factor of stability within the system: "The shifting equilibrium of forces in a multipolar balance of power encourages conciliation. . . . Anyone is a potential partner; no one is an implacable enemy" (Kegley Jr. and Raymond 1994, 51). Moreover, when noncooperative behaviors seem to condemn any progress, these advocates are a key factor to combat entrenched resistances.

As in the case of climate negotiations, the positioning of these countries consists in using the lessons one can draw from the reorganization of international power balances and the multipolarity of the international system. They are pleading the case for a change in economic models and for a collective action able to make these new development pathways operational. Some of these "middle countries" assert these orientations in their domestic policies, whereas the biggest emerging countries are making their way to environmental action with great difficulty. The latter seek to save time and to complete their economic catch-up, without really contesting their trajectory of reference, due to a lack of internal consensus that is necessary for striking a new path.

The moral and practical stand for the defense of public goods such as the environment and sustainability brings a new dimension of power to these middle countries, reflecting the great symbolic and political potential kept by these issues in international relations. The stable establishment of these new actors in the international field does not fundamentally transform the positions of major actors. However, they can—in parallel with the European action—modify the level of the obtained balance by imposing more transparent and stronger rules of the game, as well as by improving the objectives being pursued. In this sense they are the best allies of the European Union at this stage.

The Paris agreement is indeed the result of converging forces: bilateral compromises between China and the United States have set a floor. Other countries gathered around the EU and Latin American countries in the "high ambition coalition," eventually including the United States and Brazil, in an effort to support a more ambitious set of goals and more precise rules. That conjunction of forces allowed the French presidency of COP 21 to deliver an agreement that went beyond common expectations and that was qualified by the international press as a historical agreement. Compared with most international relations theories, cooperation in a multipolar world can be a serious option, a way for nations to reaffirm their sovereignty.

CONCLUSION

THE COLLISION between inevitable demographic growth and evitable albeit impending natural capital devastation will be lethal for most living organisms, humans included. This is the worst-case scenario.

Settlements will be erased inland by desertification or flooding, and on the coasts by rising seas. Water and food availability will be dramatically reduced by relentless extreme meteorological events, by water squandering and soil exhaustion, and by ocean impoverishment. Transmissible diseases will spread as temperatures increase, previous meteorological patterns change, and infectious agents lurking in tropical forests are let loose by deforestation. Hordes of desperate would-be migrants will be repelled on all sides and by all means. Among the casualties: democratic institutions and any remains of international order and moral sentiment.

But it is not impossible to change course, to make a transition to a radically different path. Science and technology are at the root of the developments that lead us to doom and they may well bring on yet other nasty surprises. However, they also provide to an amazing extent tools for making the necessary transition and, as a bonus, for lifting the global economy out of its present sluggishness—as advocated by Nicholas Stern[1]—in a distinctly more positive way than World War II ended the Great Depression.

Science and technology thus provide a powerful lever. Nature itself can help: where and when left unscathed, oceans and forests abetted by intelligent management restore the diversity of life at a remarkable pace. This will never compensate for the ongoing devastation, but when devastation is seriously contained, it will help. As far as climate change is concerned, a possible action is to pump CO_2 from the atmosphere and use it as an input in the production of alternative fuels. This will never compensate for the current volume of emissions, but it could complement a serious effort at reducing them.

The law and economics also contribute valuable tools. The law can provide regulations for natural capital use, institutions for dealing with abuses, and frameworks for the diffusion of technical innovations. Economics can provide powerful incentives to making appropriate choices pertaining to natural capital. "Man does not live on bread alone" (Deuteronomy 8:3). Indeed there will be no decisive turnaround without conscious and purposeful involvement in behavioral changes. But man does live on bread and all the goods and services that are on offer, to such an extent that significantly changing their prices deeply changes behavior.[2] Hence the paramount importance of channeling through prices, either by market or fiscal means, the values of the various components of natural capital, rewarding positive actions and penalizing damaging ones.[3] In this respect, setting internationally coordinated carbon prices is essential. This approach should nevertheless by no means be restricted to carbon: it is relevant for all significant damages to natural capital, as the Swedish experience so neatly shows.

The tool-kit for sustaining a transition is well-stocked. However that is not enough to ensure that the tools will be put to use effectively and on time. The obstacles to the transition are daunting. As numerous observations have taught Duncan Green, Oxfam senior strategic adviser, "Powerful players who stand to lose money or status from reform can be very adept at blocking it" (Green 2016).

As a private person, you may be sensitive to the state of the environment and natural capital. If you are an executive, say at a company dealing with fossil fuels, with cars, with chemicals, or in the industrial food sector, you will nevertheless act according to the logic of the firm, the logic of oil, coal, car or other products marketed, under the primacy of profit and growth. Such organizations are inevitably focused and narrow-minded. They change course only under really powerful incentives and constraints. Business models and behaviors have, for more than two centuries, crystallized around the plundering of natural capital or, as the economist Gilles Rotillon puts it, around the systematic dismissing of external effects, those "unpaid for external effects that are killing us," in Geoff Heal's words (Rotillon 2010; Heal 2016). With few exceptions yet, no firm is prepared to fundamentally change the orientation of its business model toward public interest concerns, rather than going down its own seemingly comfortable way on the common road toward the cliff. The exceptions have specific interests that may be linked to the dependence of their inputs on natural capital,[4] or to the expectations of the nature-conscious customers they serve, or they are purveyors of equipment for the transition.

Deniers and plunderers are not a specificity of market capitalism. The hybrid Chinese state firms are no better, the Soviet firms were even worse, as are the crony ones in several developing and emerging countries.[5] And, as repeatedly illustrated in this book, the more a firm's actions are harmful to both people and natural capital, the more their management is dim-witted and cynical, often hiding behind a smoke screen of actual or fabricated uncertainty.

The most celebrated pronouncement in economics, Adam Smith's on the "invisible hand"—"By pursuing his own interest he frequently promotes that of society more effectually than when he really intends to promote it" (Smith 1776)—is more often, by far, invalidated than vindicated in our economy and society.

Many large firms are wired to circumvent, cheat, or bend tax rules, health and environment regulations, public service standards, labor protections, and even basic human rights wherever and whenever they are poorly enforced by public authorities. The world is awash with cheap money emanating from central banks and from corporate profits; this money doesn't benefit the transition but rather those actors who are busy making the transition impossible. These modes of dominance dwarf those of monopoly and collusion, currently chased by economists as hindering the efficient functioning of competitive markets.[6]

Who then is to be in a position to perceive that mankind is running toward the cliff? National public authorities? To various degrees, most of them are trapped in networks of vested interests, unable to deal with unusually complex problems, particularly long-term ones, and obsessed with traditional concerns of sovereignty. When, for example, sovereign states jockey for positions in the Arctic, carving fiefdoms is the issue, not conservation. As far as public opinion is concerned there is a propensity to shrug off what are perceived as disturbing and intractable problems, expecting others (other generations?) to deal with them.

That attitude, however, is evolving as the threats become more manifest and as credible solutions emerge.[7] A multiform process of mutual inspiration and emulation and of progressive coordination of initiatives is emerging. With it original institutions—as always in this book, "institutions" is used in Douglas North's broad sense—are being built. For all its insufficiencies, the 2015 Paris Climate Agreement is one; it serves as a coordination device and a reference that guides and legitimizes steps on the transition path.

"We'll be blocking pipelines, fighting new coal mines, urging divestment from fossil fuels—trying, in short, to keep weakening the mighty industry that still stands in the way of real progress" (McKibben 2015). Bill McKibben is right: well-organized actions by groups of citizens and NGOs are essential to sustain a meaningful transition. However, they need the involvement of other actors.

Cities throughout the world, in particular coastal cities, are forming leagues to find ways of adapting to the perturbations already seen as inevitable, and of blocking those that are still avoidable. Cooperation and coordination are powerful engines to drive forward actions that isolated cities might not dare launching, for instance a ban on diesel cars, even on fossil fueled ones.

Countries in Northern Europe, provinces in Canada, and states in the western United States have chosen, or are choosing, to price carbon at significant rates. China has announced it is going down the same path. Formal mechanisms for transferring water, fishing, or building rights have been introduced in places as varied as Australia,

France, New Zealand, Oman, and Switzerland in order to far more efficiently allocate water, marine, or land resources. Many new regulations and support to green investment are implemented in a growing number of countries sending a stronger signal to economic actors.

Companies once entrenched in the intensive exploitation of natural resources are reconsidering their business model: circular economy, zero-carbon activities, sale of services instead of material goods. The long-term planning of this transition is now on the agenda of more and more businesses uncertain of the future or already convinced that a profound transformation is inevitable. We are a long way from where we were twenty years ago; but the way to get out of the trap we are still in is even longer, and the time to make it shorter.

Some movements in the financial sector point to an accelerating movement away from high carbon–footprint assets and toward assets in line with the transition to a decarbonized economy. One outcome will be—already is for certain predominantly coal assets—to make some particularly "dirty" assets so cheap that they may be acquired for liquidation.

A coalition of NGOs and other actors are currently fighting the opening in Queensland, Australia, of what would be one of the largest coal mines in the world—in what is possibly the largest untouched coal basin, the Galilee basin—as well as the construction of an accompanying dedicated port in the middle of the Great Barrier Reef Marine Park. These actors have found unexpected bedfellows: banks that were supposed to finance the operations but have in fine declined to do so, and experts at Queensland Treasury who have recommended that the public subsidies sought by the promoters of the project be refused.[8] The build-up of this de facto coalition is the most promising trait of the fight going on in Australia; it makes the implementation of the project no longer the foregone conclusion that it otherwise would have been.

Two prominent victims of the colonization of Latin America are Indian communities and rainforests. At least in the Maya Biosphere Reserve (Tikal, Guatemala) and on the lands of the Surui people (Rondonia, Amazonia, Brazil), the victims are helping each other to recover. The communities involved are earning international praise (and support) for the way sustainable conditions are obtained, in the forests they control, from a mix of ancestral knowledge and advanced techniques, properly calibrated in the spirit of Einstein's rule: As simple as possible but no simpler. Such experiences have enormous value as examples for others,[9] hence as vital contributions to the transition. It is also a mix of ancestral knowledge and modern life and earth sciences that supports the dissemination of agroecology, agroforestry in particular, as ways of sustainably feeding countless poor communities—especially in times of climate change—while regenerating soil and water resources.

In Book 3 of *L'Ancien Régime et la Révolution*, Alexis de Tocqueville shows how in the second half of the eighteenth century mushrooming new visions and actions converged to overthrow the Old Regime and generate a new order. Here there is a similar pattern with the perspective of an equally broad revolution. Famously, and unwisely, President George Bush the elder, rejecting any serious perspective of transition, declared at the Rio Earth Summit in 1992: "The American way of life is not negotiable." In reality there is no way people in rich countries can preserve a certain quality of life without in-depth changes in their way of life. And in other countries there is no way people can access a certain quality of life if they aim at the American way of life.

In his *Remarks to the United Nations General Assembly*, in September 2015, Pope Francis declared: "To enable these real men and women to escape from extreme poverty, we must allow them to be dignified agents of their own destiny." Economists tend to distinguish equity concerns from efficiency concerns. Pope Francis deals with both together: escaping from poverty without further jeopardizing natural capital is possible only if the people involved devise and implement their own way on the transition path. Support from the outside is both necessary and welcome; dictating conduct is not.

In a forceful book where he presents broad and practical policy proposals aiming at reducing inequality in contemporary societies, Tony Atkinson writes: "The world faces great problems, but collectively we are not helpless in the face of forces outside our control. The future is very much in our hands" (Atkinson 2015). What a fitting conclusion for the present book as well, under a stringent condition: no more time can be wasted. Sensitive to climate change as much as to nuclear threats, the long arm of the Clock in the Bulletin of Atomic Scientists is at two and a half minutes to midnight (from three minutes in 2016: a Trump effect), "marking the direst setting of the Clock since 1983, at the height of the Cold War." This symbol of urgency should be taken very seriously: life on earth is on the brink. Time to pass the baton to women, men having so blatantly failed?

NOTES

INTRODUCTION

1. We use the term *institution* in Douglass North's broad sense, as in his Nobel Lecture in 1993: "Institutions are the humanly devised constraints that structure human interaction. They are made up of formal constraints (rules, laws, constitutions), informal constraints (norms of behavior, conventions, and self-imposed codes of conduct), and their enforcement characteristics."

1. EROSION OF BIOLOGICAL DIVERSITY

1. Among the empirical studies, Newbold et al. (2016) is particularly broad. The editors of *Science* summarized the authors' method for defining the limits of biodiversity loss: "Using over 2 million records for nearly 40,000 terrestrial species, they modeled the response of biodiversity to land use and related pressures and then estimated, at a spatial resolution of ~1 km^2, the extent and spatial patterns of changes in local biodiversity. Across 65 percent of the terrestrial surface, land use and related pressures have caused biotic intactness to decline beyond 10 percent, the proposed 'safe' planetary boundary. Changes have been most pronounced in grassland biomes and biodiversity hotspots."
2. Not including water pollution and methane leaks (see chapter 3).
3. Small quantities of this sand have even been detected in Southern California.
4. For a broader account of the importance of insects to the environment and human activities, see Goulson (2014), who superbly illustrates a remark

by the famous biologist Edward O. Wilson, dubbed "the father of biodiversity": "If insects were to vanish, the environment would collapse into chaos."

5. The title of Bagley et al. (2014) is formulated as a question: "Drought and Deforestation: Has Land Cover Change Influenced Recent Precipitations Extremes in the Amazon?" The answer that emerges from the broad investigation reported in the article is an unambiguous yes. The huge effort by a consortium of more than 150 scientists at "estimating the global status of more than 15,000 Amazonian tree species" brings the conclusion that "at least 36 percent and up to 57 percent of all Amazonian tree species are likely to qualify as globally threatened under IUCN [International Union for Conservation of Nature] Red List criteria."
6. See the Annual Review of the International Tropical Timber Organization, www.itto.int/annual_review/.
7. Fish-aggregating devices (FADs) are drifting or moored. Even when they are as simple as bamboo rafts (which, by the way, often use GPS localization), they are highly efficient, as fish tend to gather in great numbers around floating objects. According to Davies et al. (2016), FADs help catch a million tons of tuna annually, with sharks and marlins often being undesirable bycatch.
8. See Jim Yardley, "Two Hungry Nations Collide Over Fishing," *New York Times*, September 4, 2012.
9. See Timothy Allen, "Pa aling Fishing," http://humanplanet.com/timothyallen/2011/01/pa-aling-fishing/.
10. Anthony Ricciardi is associate professor at the Department of Biology, McGill University, Montreal, Canada.
11. The fifth mass extinction, which occurred 65 million years ago, entailed the disappearance of the dinosaurs, as well as many other animals and plants. For past mass extinctions and the perspectives of a sixth one, see Kolbert (2014).

2. THE UBIQUITOUS WASTE AND GROWING SCARCITY OF WATER AND SOIL

1. The realization of one's capacities, in Amartya Sen's (1983) terms.
2. From 2002 to 2010, Anna Tibaijuka was under-secretary general of the United Nations and executive director of UN Human Settlements Programme (UN-HABITAT).
3. Regulations are also part of water management in Nevada. Lawns in front of houses are prohibited. Golf courses and casinos must lay down specific networks for nondrinkable water for watering their grass or for feeding their fountains and cascades.

4. See "The Price of Water," Circle of Blue, http://www.circleofblue.org/waterpricing/.
5. We are deeply indebted to Daniel Hillel for the insights drawn from his considerable achievements in soil and water science and management.
6. Trees support agriculture in innumerable ways. A remarkable study published in *Ecology Letters* (Karp et al. 2013) measured the sensitivity of coffee yields in Costa Rica to the existence of forest patches in the coffee plantations. The trees had no direct effect on the coffee plants; however, they did provide shelter for warblers and wrens, small birds that feed on berry borer beetles, which are the most damaging insect pests on coffee. The authors completed their study by estimating, in dollars, the savings obtained in this way; it appears that the gain per hectare is roughly equal to the average individual income in Costa Rica.
7. At low pH values (pH < 5.5), the toxic ion of aluminum (Al3+) is solubilized from aluminosilicate clay minerals into soil solutions and is toxic to crop plants. Aluminum toxicity mainly targets the root apex, resulting in inhibited root growth and function. As a result, aluminum toxicity leads to severe impairment in the plant's acquisition of water and nutrients from the soil, which results in a significant reduction in crop yields on acid soils. Up to 50 percent of the world's potentially arable soils are acidic, with a significant proportion of these acidic soils found in the tropics and subtropics in developing countries, where food security is most at risk. Thus, aluminum stress represents one of the most important constraints for agricultural production worldwide (Royal Society 2009).
8. Farmers in France have been granted financial compensation not only for various cancers but also for Parkinson disease induced by the repeated manipulations of pesticides.
9. Confiscated pieces of land are subsequently used to build resorts aimed at well-off Chinese living across the border.
10. For two valuable references dealing with these sensitive issues, see Nayar (2012) and Worldwatch Institute (2015).

3. ENERGY: AS LITTLE AS POSSIBLE

1. On dangers associated with raw wood burning, see chapter 8.
2. On top of the CO_2 budget, there is room for up to 50 percent of this budget in emissions of other greenhouse gases—in particular, methane—converted into CO_2 equivalents. According to Winkelmann et al. (2015), one consequence of burning the totality of proven fossil fuels would be the complete elimination of the Antarctic ice sheet, which "stores water equivalent to 58 m in global sea-level rise" (1). This seems to invalidate the confidence expressed by the chief executive officer of Exxon-Mobil and his

counterpart at Chevron that they will be able to make use of all the reserves controlled by their respective companies.

3. This viewpoint is vindicated in the March 2012 Carbon Tracker Initiative.
4. Remarks by World Bank Group president, Jing Yong Kim, at the World Bank Group–International Monetary Fund Annual 2016 Climate Ministerial.
5. Alvarez et al. (2012) estimated at 3.2 percent "the threshold beyond which gas becomes worse for the climate than coal" (6437). For assessments of actual leaks in the United States, see Allen et al. (2013) and Marchese et al. (2015).
6. No less than 25,450 gas and oil wells have been evaluated by Monika Freyman and her team at CERES in the course of preparing this report (Freyman and Salmon 2013).
7. From 1978 to 2008, there was an average of two earthquakes of magnitude 3.0 or greater per year in Oklahoma. From the beginning of 2009 to the end of 2015, there were sixty earthquakes of magnitude 4.0 or greater, as well as about 1,700 lesser ones. These quakes don't seem to be linked to core fracking operations but, rather, to routine disposal of huge quantities of wastewater into thousands of wells that reach underground faults that they, so to say, lubricate. Oklahoma has become the most seismically active U.S. state.
8. As Joskow (2013) mentioned, "Over the next decades hundreds of thousands of new shale gas wells are expected to be drilled across the United States."
9. In Massachusetts, they experiment with a way to regulate the use of the bridge: the Massachusetts Energy Board has approved the project of a gas-fired power plant in Salem Harbor under conditions of progressive reduction of CO_2 emissions and closure by 2050 at the latest.
10. For a broad, balanced assessment of the perspectives of civil nuclear power, see Lévêque (2014).

4. PERSPECTIVES ON CLIMATE CHANGE

1. On February 8, 1965, President Lyndon Johnson delivered a special message to Congress on conservation and restoration of natural beauty that includes a paragraph raising the issue of climate change: "Air pollution is no longer confined to isolated places. This generation has altered the composition of the atmosphere on a global scale through radioactive materials and a steady increase in carbon dioxide from the burning of fossil fuels." That same year, Roger Revelle chaired a subcommittee of the president's Science Advisory Committee, which wrote a chapter on global warming in the committee's *Restoring the Quality of Our Environment* report.

2. For a systematic review, see the *IPCC Fifth Assessment Report, 2014*. From the IPCC website: "The Intergovernmental Panel on Climate Change (IPCC) is the leading international body for the assessment of climate change. It was established by the United Nations Environment Programme (UNEP) and the World Meteorological Organization (WMO) in 1988 to provide the world with a clear scientific view on the current state of knowledge in climate change and its potential environmental and socio-economic impacts."
3. The jet stream effect has been compared with the effect of leaving a refrigerator door open, with cold air flooding the kitchen as the interior of the refrigerator warms.
4. On October 22, 2016, Dr. Hamilton died in a fall down a deep crevasse in Antarctica.
5. For a succinct explanation of how glaciers regulate water flows, see Jansson, Hock, and Schneider (2012).
6. This inventory was completed at the Cold and Arid Regions Environmental and Engineering Research Institute of the Chinese Academy of Sciences (Guo et al. 2014).
7. The effects have been all the more devastating because, in some places, the reefs are also victims of specific aggressions. A well-known case is observed along the Great Barrier Reef, off the eastern coast of Australia: the aggressor is the crown-of-thorns starfish, which emits enzymes dissolving the coral into an edible soup. During the past thirty years, populations of the starfish have repeatedly surged on the Great Barrier Reef, and these surges appear to be correlated with episodes of pollution. In the Caribbean Sea, overfishing of parrotfish has deprived coral reefs of an essential consumer of algae, compromising the ecological balance between corals and algae (see chapter 1). In the Great Barrier Reef, another major threat is coal mining and moving along the coastline, potentially on a scale never seen anywhere before (see the Introduction).
8. Of particular concern are the beetles that cut off the path within the tree between soil and needles—hence, between nutrients and chlorophyll. These beetles survive in greater number during milder winters.
9. For a broad review of the condition of world forests, see the special issue of *Nature Climate Change*, August 21, 2015.
10. This is not to say that the western United States is comfortable. In Phoenix, Arizona, for example, temperatures in July 2016 were never below 45°C, and the only possible way to adapt was to stay indoors (house or car) during the day, getting out only in the early morning. Extreme heat is detrimental in many ways. It has recently been shown to systematically increase the risks of conflicts. Concluding a broad investigation, Burke, Hsiang, and Miguel (2015a) wrote: "Adverse climatic events increase the risk of violence and conflict, at both the interpersonal level and the intergroup level, in

societies around the world and throughout history. . . . We find that temperature has the largest average effect by far" (610).

See also the paper published in *Nature* by the same authors (2015b).

The effects of heat waves would be even more horrific in Southeast Asia (Im, Pal, and Eltahir 2017).

11. Remarks delivered by President Obama at the 2016 Pacific Islands Conference of Leaders, Honolulu, September 1, 2016.

5.ENLISTING THE SCIENTIFIC METHOD

1. It is actually not clear whether Galileo ever performed experiments on gravity from the top of the tower.
2. For a detailed account, see Riordan and Hoddeson (1997).
3. Harold Kroto himself dealt with the issue in Kroto (1997); see also chapter 7.

6. SUSTAINABILITY AT THE INTERSECTION OF SCIENCE AND NATURE

1. Halpern and Warner (2002, 361), after reviewing eighty marine reserves aged from one to forty years, came to the following conclusion: "Biological responses [in terms of density, diversity, and size of organisms—fishes, in particular] inside marine reserves appear to develop quickly and last through time" (for a more recent review, see Innis and Simcock 2016). Coral Castles in the Phoenix Islands of the Pacific offer an example of a coral reef that was considered dead in 2003 but was found to have come back to multicolor life in 2015 (Witting 2016). In the intervening years, the government of Kiribati (a group of tropical islands, including the Phoenix Islands, halfway between Hawaii and Australia) had created the Phoenix Islands Protected Area (PIPA), the largest such area in the Pacific, with more than 400,000 km^2. Thanks to this effort, PIPA—and, in particular, Coral Castles—is no longer exposed to damages caused by local human activities. Even more recently, introduced damaging alien species have been eradicated. The region's UNESCO World Heritage recognition testifies to the significance of the results obtained by the Kiribati pioneers. The question remains: Will these results withstand the relentless warming and acidification of the ocean, which, at the level of Kiribati, are beyond any possible action?
2. SWP adds: "We will lose the opportunity [to remove atmospheric CO_2] if society delays research and development to lower the technical barriers to efficacy and affordability of carbon dioxide removal." See also Geden and Schäfer (2016).

3. In particular, in Andra Pradesh, Gujarat, Madhya Pradesh, and Maharashtra. See Dinesh C. Sharma, "Bt Cotton Has Failed Admits Monsanto," *India Today*, March 6, 2010.
4. There are still difficulties in completely eliminating some pharmaceutical residues.
5. Orange County Water District, "Awards." https://www.ocwd.com/news-events/awards/. Accessed August 7, 2017.
6. For detailed information, see *Precision Agriculture*, a journal published by Springer since 1999.
7. Reduction rates of up to 75 percent are reported in Mortensen et al. (2012).
8. The role of perennials in improving agricultural practices in Africa is assessed in Glover, Reganold, and Cox (2012).
9. For a detailed review of solid-state batteries, either lithium or non-lithium based, see Kim et al. (2015).
10. As are many pieces of equipment essential to a decarbonized economy (see Abraham 2015).
11. Fraas (2014) provided a reliable and readable account of the history, diversity, and perspectives of photovoltaic cells.
12. Albeit, not the misguided policies aiming at commercial dissemination of outmoded equipment.
13. Heat pumps warm or cool air in buildings by taking advantage of the earth's nearly constant underground temperature.
14. As a clear illustration, one can mention RedBus, the Indian IT start-up that coordinates thousands of long-distance bus lines run by independent companies, many rather small, all over the country. It continuously disseminates information about available services to all bus lines and to consumers.
15. Center for Negative Carbon Emissions (Arizona State University); Carbon Engineering (Cambridge, Mass., and Calgary); Global Thermostat (New York and Princeton, N.J.); Climeworks (Zürich); Combustion and Carbon Capture Technologies (Chalmers University of Technology, Gothenburg, Sweden).
16. At the Loker Hydrocarbon Research Institute, University of Southern California.

7. SCIENTIFIC UNCERTAINTY, FABRICATED UNCERTAINTY, AND THE VULNERABILITY OF REGULATION

1. "Much of the time, we function in an ambiguous zone, without clear-cut answers," as Danielle Ofri, professor at New York University School of Medicine, wrote in "Uncertainty Is Hard for Doctors," *New York Times*, June 6, 2013.

2. He subsequently made rapid progress to more complete theoretical and experimental foundations and became a Nobel laureate in 1997, which was rather soon after the publication of his results.
3. David Goodstein is professor of physics at the California Institute of Technology; he has been vice provost in charge, among other things, of checking scientific misconduct. His book *On Fact and Fraud* (Goodstein 2010) offers a selection of significant cases of scientific fraud.
4. For a comprehensive assessment of the damages to the development of the brain, see Grandjean (2015).
5. Horel (2015) offered a well-informed and brilliant book about how the European Commission dealt with endocrine disruptors.
6. Atrazine is an insecticide applied to more than half the U.S. corn crop. It is produced by the Swiss firm Syngenta. It has been banned in the European Union for dangerously polluting drinking water. From then on, the U.S. market has been by far the largest one.

8. PRODUCING AND DISSEMINATING SUSTAINABILITY-ENHANCING INNOVATIONS

1. The initial formulation is in Albert Einstein's Herbert Spencer Lecture (1933): "It can scarcely be denied that the supreme goal of all theory is to make the irreducible basic elements as simple and as few as possible without having to surrender the adequate representation of a single datum of experience" (*Philosophy of Science* 1[2]: 163–169, 165). It evolved to "Everything should be made as simple as it can, but no simpler" as quoted (and applied to music) by composer Roger Sessions in an article published in the *New York Times* (January 8, 1950, "How a Difficult Composer Gets That Way") and finally to "Make it as simple as possible, but no simpler."
2. According to the FAO, drylands are characterized by water deficits—that is, by unfavorable balances between water precipitation and evapotranspiration (0.65 on average). Drylands cover about 45 percent of the world's land surface and are occupied by a third of the world's population, with a high proportion of very poor households.
3. As is shown in Shapiro (2000), who considers how to "navigate the patent thicket."
4. According to traditional principles, a discovery is not patentable; only an invention is. Nevertheless, for about the past thirty years, the main patent offices and the courts have ignored this.
5. "Market power is not to be presumed" means that not all patents automatically create problems from the point of view of competition protection; however, problems (possibly serious ones) derive from the absence of close substitutes and thus need remedies. For an extremely well-documented

report on the relationships between competition policy and the protection of intellectual property rights, see U.S. Federal Trade Commission (2003).

6. At the time, Robert Barr was vice president for intellectual property of Cisco Systems, before returning to Berkeley School of Law (University of California).
7. For a more detailed account, see Henry and Stiglitz (2010).
8. The Uruguay Round started in 1982 and did not conclude until 1994.
9. Robert C. Pozen was chair of MFS Investment Management after having served as vice chair of Fidelity Investments.
10. Revelation mechanisms play a central role in contemporary economic analysis, as shown in Eric Maskin's Nobel lecture (2008).
11. John Barton was an engineer and U.S. Navy officer turned lawyer who became professor at Stanford Law School; there he specialized in topics at the intersection of science, law, and society—in particular, innovation and intellectual property. He was concerned with promoting responsible use of natural capital. In 2009, he died from a brain injury sustained in a bicycle accident in Los Altos, California.
12. See also Boldrin and Levine (2004, 2008), who identified several industries, such as chemical and software, that at critical stages developed without IPR protection.
13. As illustrated in von Hippel (2005).
14. Established in 1843 and known for a long time as Rothamsted Experimental Station.
15. Chemical communication between plants and between plants and animals is not a rarity. "Common mycelial networks can act as conduits for signaling networks" (Babikova et al. 2013, 835)—for example, between beanstalks, which then chemically repel enemies and attract potential allies. When a beanstalk is attacked by voracious aphids, it warns neighbors by sending specific molecules along the network of filament fungi that connect the plants underground. Upon reception, the neighbors emit volatile chemicals that not only tend to repel aphids but also attract aphid-eating wasps.
16. Sarah Toumi was born in France to a French mother and a Tunisian father, an engineer who used to work in nonprofit organizations. After graduating from the University of Paris, she settled in the village where her grandparents were living, married a farmer, and created "Acacias pour Tous" ("Acacias for All") as a vehicle to promote agroforestry within communities, like her village, threatened by the combined effects of climate change and desert expansion. She was granted several prizes for her action and in 2016 appeared on the Forbes list of "30 Under 30 Social Entrepreneurs."
17. UN Food and Agriculture Organization, "Traditional Crop of the Month: Moringa," http://www.fao.org/traditional-crops/moringa/en/.

18. Interesting examples are found in the Amazon basin, where the soils are relatively poor in the nutrients that cultivated crops require. There are, however, pockets of fertility that have been created by ancient Indian communities when they mixed natural soils with biochar and organic detritus.

9. ECONOMIC INSTRUMENTS FOR SUSTAINABLE DEVELOPMENT

1. Originally published in 1759, the book stemmed from Smith's multidisciplinary teachings at the University of Glasgow.
2. For a concise introduction to the economics of incentives, see Dasgupta (2007), chapters 4 and 7. For an extended review of environmental taxation, see Milne and Andersen (2012).
3. When informing Prime Minister Margaret Thatcher of such balancing shifts, British economist David Pearce coined the expression "double dividend."
4. From 1997 onward, the floor has been down at 25 MWh. The proceeds are refunded to the plants in relation to their respective fractions of total useful energy produced; the refunding formula doesn't interfere with the incentives generated by the tax mechanism. See Organisation for Economic Co-operation and Development (2010).
5. S 98, 105th Cong., https://www.congress.gov/bill/105th-congress/senate-resolution/98/text.
6. The phasing out of fossil fuel use for heating is well on its way, already reaching 100 percent for residential buildings and 70 percent for the service sector, shops, offices, and so forth. The prevalent heating fuels are now wood pellets and waste, which burn all the more efficiently because district heating is dominant in urban areas.
7. Swedish Environmental Protection Agency, "The Generational Goal," August 29, 2016, http://www.swedishepa.se/Environmental-objectives-and-cooperation/Swedens-environmental-objectives/The-generational-goal/.
8. However, the titles of two op-eds by *New York Times* columnist Thomas L. Friedman sound like a call to arms: "The Market and Mother Nature" (January 8, 2013) and "It's Lose-Lose vs. Win-Win-Win-Win-Win" (March 16, 2013). Also on March 16, 2013, the main *Financial Times* editorial read: "Now Is Right Time for Least-Worst Tax: Carbon Levy Is a Better Way to Raise Revenue." This idea can be made operational as shown in Dinan (2012), Metcalf (2009), and Metcalf and Weisbach (2009).
9. All sorts of pollution are becoming unbearable in China. However, as Elizabeth Economy (2004) forcefully showed, polluting is such a significant component to profit making by so many members of the Communist Party, it requires a political and economic revolution to meaningfully reduce it.

10. The Financial Stability Board is an international institution created following a decision by the G20; it is hosted in Basel by the Bank for International Settlements.

10. GLOBAL GOVERNANCE OF SUSTAINABLE DEVELOPMENT

1. "Overview of the Millennium Ecosystem Assessment" can be found at http://www.unep.org/maweb/en/About.aspx.
2. See Scholte (2002, 295–298) for a review of these challenges.
3. The Rio Dialogues website can be found at http://www.riodialogues.org.
4. The results of the final vote, including disaggregated data by continent, by human development index (HDI), by age, and by gender, are available at http://vote.riodialogues.org.

11. THE GEOPOLITICS OF ENVIRONMENT

1. To mention only the most famous one, see Rosenau (1995).
2. For the complete inaugural address, go to https://obamawhitehouse.archives.gov/blog/2009/01/21/president-barack-obamas-inaugural-address.
3. "Beijing's levels of PM2.5s—particles that are smaller than 2.5 micrometers in diameter and can penetrate the gas exchange regions of the lungs—are the worst in the world. Beijing's 2012 March average reading was 469 micrograms of such particles per cubic meter, which compares abysmally with Los Angeles' highest 2012 reading of 43 micrograms per cubic meter" (T. N. Thompson, "Choking on China," *Foreign Affairs*, April 2013).
4. "Safeguarding Environment a Priority," *China Daily*, November 29, 2010.

12. THE NEW MULTIPOLARITY OF SUSTAINABLE DEVELOPMENT

1. John Vogler and Hannes Stephan (2007) indicated that the EU has developed policy competencies across a range of environmental areas and has signed more than sixty multilateral environmental agreements.
2. For climate negotiations only: BASIC (Brazil, South Africa, India, and China); the least developed countries (LDCs), consisting of forty-five of the world's poorest nations, mostly in Africa; the African Group; the Oil Producing and Exporting Countries (OPEC); the Arab States, mostly in OPEC, but some not; the Association of Small Island States (AOSIS), which also includes Bangladesh and some countries not in the G77 (totaling forty-two

member-states and observers); Small Island Developing States (SIDS), which has a different membership from AOSIS); ALBA (Bolivarian Alliance for the Peoples of Our America), which includes Cuba, Venezuela, Bolivia, Nicaragua, Honduras, Dominica, and Saint Vincent and the Grenadines, among others; the Central American Integration System (SICA); The Group of Mountain Landlocked Developing Countries (Armenia, Kyrgyzstan, and Tajikistan); Central Asia, Caucasus, Albania, and Moldova (CACAM); the Coalition for Rainforest Nations; the Cartagena group (forty developed and developing countries); the Environmental Integrity Group (EIG), which consists of Mexico, Liechtenstein, Monaco, the Republic of Korea, and Switzerland; the Independent Alliance of Latin America and Caribbean (AILAC); and the Like-Minded Group of Developing Countries (LMDC).

CONCLUSION

1. In particular in his latest book: *Why Are We Waiting? The Logic, Urgency, and Promise of Tackling Climate Change*, 2015.
2. Here is a limited albeit emblematic example. It is discussed by Edward Luce in "SUVs cannot lead the climate change convoy," *Financial Times*, November 23, 2015: as the price of gasoline came down during the 2014 summer and on, SUVs and light trucks sales picked up (more than 20% In 2015), while sales of hybrid cars decreased dramatically.
3. Elder statesmen James A. Baker III, Henry M. Paulson Jr, and George P. Shultz tried to convince fellow Republicans that taxing carbon is an efficient and fair (with proceeds being returned to households in ways that don't interfere with the incentive to reduce fossil fuel consumption) instrument to deal with an all-too-real-climate change.
4. According to Jeffrey Seabright, chief sustainability officer at Unilever, formerly vice president of environment and water resources at Coca-Cola: "Increased droughts, more unpredictable variability, 100-year floods every two years, when we look at our most essential ingredients, we see those events as threats" (*New York Tim*es, January 23, 2014).
5. Consider three defunct lakes. The Aral sea was the largest inland water body in Asia; it dried up courtesy of massive Soviet industrialized agriculture. Lake Urmia was to Iran what the Great Salt Lake is to the United States; it has been plundered by enterprises controlled by the Islamic Revolutionary Guards Corps. Lake Poopo, in Bolivia, second largest lake in the Andes after Lake Titicaca, was sacrificed to the development of mining activities supported by President Morales, despite the devastation inflicted on Indian communities on the Andean plateau. In the cases of Lake Urmia and Lake Poopo, climate change is also seen as a factor of desiccation.

6. Indeed: Chemical and food companies extensively harm health and environment and sabotage attempts at regulating their damaging activities (see chapters 1, 2, and 7).

   Providers of fossil fuels and industries hinging on them deploy their huge power and resources against the vitally necessary transition to a low carbon economy (see chapters 3 and 7).

   Mining companies devastate lives and environment in poor countries, which they push to inequality, corruption, mafia rule and even civil war (see introduction and chapter 7).

   Pharmaceuticals push toward excessive consumption where there is money to pay for the drugs they sell, and make any effort to restrict their availability where there is little money to make (see chapter 8).

   Before branding this assessment excessive, think of these companies' enthusiastic reception of the dash for deregulation pledged by President Trump. Would responsible actors of an indispensable transition react in this way?

   IT companies have not played a prominent role in this book. It is nevertheless worth reminding that they are busy searching for ways to make most people useless in production processes, in both industry and services, while a tiny minority concentrates knowledge, power and wealth, hence maximizing inequality and the chances of major disruptions in society. Few people are as lucid about these perspectives as is Stephen Hawking: "The automation of factories has already decimated jobs in traditional manufacturing, and the rise of artificial intelligence is likely to extend this job destruction deep into the middle classes, with only the most caring, creative or supervisory roles remaining. This in turn will accelerate the already widening economic inequality around the world. The Internet and the platforms that it makes possible allow very small groups of individuals to make enormous profits while employing very few people." ("This Is the Most Dangerous Time for Our Planet," *Guardian*, December 1, 2016). "Widening economic inequality," beyond the present levels where the top eight men (no women in this small group) own the same wealth as the bottom 3.6 billion people (Oxfam, *An Economy for the 99 Percent*, Oxford, 2017), would entail an irreversible expansion of mass poverty, hence of natural capital destruction.
7. Robert Jay Lifton, Harvard psychologist and historian, sees a possible shift in the depths of American minds and souls: "Americans appear to be undergoing a significant psychological shift in our relation to global warming. I call this shift a climate 'swerve,' borrowing the term used recently by the Harvard humanities professor Stephen Greenblatt to describe a major historical change in consciousness that is neither predictable nor orderly. . . . Could the climate swerve come to include a 'climate freeze,'

defined by a transnational demand for cutting back on carbon emissions in steps that could be systematically outlined?" (*New York Times*, August 23, 2014).

8. Also, despite large coal resources from which it can tap, AGL Energy, Australia's biggest utility, is progressively substituting large solar farms for coal-fired power stations.
9. Naomi Klein, in *This Changes Everything: Capitalism Versus the Climate* (2014), reviews numerous conflicts pitting indigenous communities against private or public actors whose endeavors are major threats for the natural capital on which these communities depend.

REFERENCES

Abraham, David. S. 2015. *The Elements of Power: Gadgets, Guns, and the Struggle for a Sustainable Future in the Rare Metal Age*. New Haven, CT: Yale University Press.

Afionis, Stavros. 2010. "The European Union as a Negotiator in the International Climate Change Regime." *International Environmental Agreements* 11 (4): 341–60.

Akbar, Bisma. 2011. "Pakistan's Forgotten Emergency" (press release). Oxfam International. https://www.oxfam.org/en/pressroom/pressreleases/2011-11-30/pakistans-forgotten-emergency.

Akerlof, George A. 1970. "The Market for Lemons: Quality Uncertainty and the Market Mechanism." *Quarterly Journal of Economics* 84 (3): 488–500.

Alexievich, Svetlana. "On the Battle Lost" (Nobel Lecture). https://www.nobelprize.org/nobel_prizes/literature/laureates/2015/alexievich-lecture.html.

Alexis de Tocqueville Institution. 1994. *Science, Economics, and Environmental Policy: A Critical Examination*. Arlington, VA: Alexis de Tocqueville Institution.

Allen, David T., Vincent M. Torres, James Thomas, David W. Sullivan, Matthew Harrison, Al Hendler, Scott C. Herndon, et al. 2013. "Measurements of Methane Emissions of Natural Gas Production Sites in the United States." *Proceedings of the National Academy of Sciences*, 110 (44): 17768–73. doi: 10.1073/pnas.1304880110.

Allen, Myles. 2015. "Paris Emissions Cuts Aren't Enough—We'll Have to Put Carbon Back in the Ground." *The Conversation,* December 12. https://theconversation.com/paris-emissions-cuts-arent-enough-well-have-to-put-carbon-back-in-the-ground-52175.

Alroy, John. 2015. "Current Extinction Rates of Reptiles and Amphibians." *Proceedings of the National Academy of Sciences* 112 (42): 13003–8.

Alscher, Stefan. 2010. "Environmental Factors in Mexican Migration: The Case of Chiapas and Tlaxcala." In *Environment, forced migration and social vulnerability*, ed. T. Afifi and J. Jäger, 171–85. Heidelberg: Springer Verlag.

Alvarez, Ramón A, Stephen W. Pacala, James J. Winebrake, William L. Chameides, and Steven P. Hamburg. 2012. "Greater Focus Needed on Methane Leakage from Natural Gas Infrastructure." *Proceedings of the National Academy of Sciences* 109 (17): 6435–40.

Andersson, Mats, Patrick Bolton, and Frédéric Samama. 2014. "Hedging Climate Change." Columbia Business School, Columbia University, New York. https://wwwo.gsb.columbia.edu/faculty/pbolton/papers/Hedgingclimaterisk(v35).pdf.

Antweiler, Werner, Brian R. Copeland, and M. Scott Taylor. 2001. "Is Free Trade Good for the Environment?" *American Economic Review* 91 (4): 877–908.

Arkema, Katie K., Greg Guannel, Gregory Verutes, Spencer A. Wood, Anne Guerry, Mary Ruckelshaus, Peter Kareiva, Martin Lacayo, and Jessica M. Silver. 2013. "Coastal Habitats Shield People and Property from Sea-Level Rises and Storms." *Nature Climate Change* 3: 913–18.

Arrow, Kenneth J., and Leonid Hurwicz. 1997. "An Optimality Criterion for Decision-Making under Ignorance." In *Studies in Resource Allocation Processes*, ed. Kenneth Arrow and Leonid Hurwicz, 463–71. Cambridge, UK: Cambridge University Press.

Atkinson, Anthony B. 2015. *Inequality: What Can Be Done?* Cambridge: Harvard University Press.

Atkinson, S. 2017. "A Perfect Storm in the Gulf of Alaska: Factors Contributing to 2015–2016 Common Murre Die-Off." Paper presented at the Alaska Marine Science Symposium, Anchorage, January 24, 2017.

Attina, Teresa M., Russ Hauser, Sheela Sathyanarayana, Patricia A. Hunt, Jean-Pierre Bourguignon, John Peterson Myers, Joseph DiGangi, R. Thomas Zoeller, and Leonardo Trasande. 2016. "Exposure to Endocrine-Disrupting Chemicals in the USA: A Population-Based Disease Burden and Cost Analysis." *The Lancet Diabetes and Endocrinology* 4 (12): 996–1003. doi: 10.1016/S2213-8587(16)30275-3.

Austen, Ian. 2013. "A Black Mound of Canadian Oil Waste Is Rising Over Detroit." *New York Times,* May 17.

Aviv, Rachel. 2014. "A Valuable Reputation." *The New Yorker,* February 10, 52–63.

Babikova, Zdenka, Lucy Gilbert, Toby J. A. Bruce, Michael Birkett, John C. Caulfield, Christine Woodcock, John A. Pickett, and David Johnson. 2013. "Underground Signals Carried Through Common Mycelial Networks Warn Neighboring Plants of Aphid Attack." *Ecology Letters* 16 (7): 835–43. doi: 10.1111/ele.12115.

Bäckstrand, Karin. 2006. "Multi-Stakeholder Partnerships for Sustainable Development: Rethinking Legitimacy, Accountability, and Effectiveness."

European Environment (Special Issue: Rules for the Environment; Reconsidering Authority in Global Environmental Governance) 16 (5): 290–306.

Bagley, Justin E., Ankur R. Desai, Keith J. Harding, Peter K. Snyder, and Jonathan A. Foley. 2014. "Drought and Deforestation: Has Land Cover Change Influenced Recent Precipitation Extremes in the Amazon?" *Journal of Climate* 27: 345–61. doi: 10.1175/JCLI-D-12-00369.1.

Bals, Christophe, Charlotte Cuntz, Oldag Caspar, and Jan Burck. 2013. "The End of EU Climate Leadership." Germanwatch. https://germanwatch.org/en/7563.

Barnosky et al. (2011), "Has the Earth's Sixth Mass Extinction Already Arrived?" *Nature*, 471: 51–57.

Barrangou, Rodolphe. 2014. "Cas9 Targeting and the CRISPR Revolution." *Science* 344 (6185): 707–8.

Barrett, Scott. 1994. "Self-Enforcing International Environmental Agreements." *Oxford Economic Papers* 46: 878–94.

——. 2010. *Why Cooperate? The Incentive to Supply Global Public Goods*. Oxford, UK: Oxford University Press.

——. 2012. "Climate Treaties and Backstop Technologies." *CESifo Economic Studies* 58 (1): 31–48. doi: 10.1093/cesifo/ifr034.

Barton, John H. 2007. "Intellectual Property and Access to Clean Energy Technologies in Developing Countries: An Analysis of Solar Photovoltaic, Biofuel, and Wind Technologies." ICTSD Issue Paper No. 2. Geneva: International Centre for Trade and Sustainable Development. http://www.iprsonline.org/New%202009/CC%20Barton.pdf.

Battacharyya, Subhes C. 2014. "Viability of Off-Grid Electricity Supply Using Rice Husk: A Case Study from South Asia." *Biomass and Bioenergy* 68: 44–54.

Baylon, Caroline, David Livingstone, and Roger Brunt. 2015. *Cyber Security at Civil Nuclear Facilities: Understanding the Risks* (Chatham House Report). London: Royal Institute of International Affairs.

Bee Informed Partnership. 2015. "2014–2015 National Colony Loss and National Management Survey." https://beeinformed.org/2014-2015-colony-loss-and-national-management-survey/.

Beketov, Mikhail A., Ben J. Kefford, Ralf B. Schäfer, and Matthias Liess. 2013. "Pesticides Reduce Regional Biodiversity of Stream Invertebrates." *Proceedings of the National Academy of Sciences* 110 (27): 11039–43.

Bellanger, Martine, Barbara Demeneix, Philippe Grandjean, R. Thomas Zoeller, and Leonardo Trasande. 2015. "Neurobehavioral Deficits, Diseases, and Associated Costs of Exposure to Endocrine Disrupting Chemicals in the European Union." *The Journal of Clinical Endocrinology and Metabolism* 100 (4): 1256–66.

Bertini, Catherine. 2014. "Effective, Efficient, Ethical Solutions to Feeding 9 Billion People: Invest in Women." Presented at the Crawford Fund, Canberra, Australia, August 27.

Bessen, James, and Eric Maskin. 2009. "Sequential Innovation, Patents, and Imitation." *RAND Journal of Economics* 40 (4): 611–35.

Betsill, Michelle M., and Elisabeth Corell, eds. 2007. *NGO Diplomacy: The Influence of Nongovernmental Organizations in International Environmental Negotiations*. Cambridge, MA: MIT Press.

Biermann, Frank, and Philipp Pattberg, eds. 2012. *Global Environmental Governance Reconsidered*. Cambridge, MA: MIT Press.

Bloomberg New Energy Finance. 2016. "New Energy Outlook 2016." https://about.newenergyfinance.com/international/china/new-energy-outlook/

Boldrin, Michele, and David K. Levine. 2004. "The Economics of Ideas and Intellectual Property." *Proceedings of the National Academy of Sciences* 102 (4): 1253–56.

——. 2008. *Against Intellectual Monopoly*. New York: Cambridge University Press.

Bolton, Patrick, Frederic Samama, and Joseph E. Stiglitz, eds. 2011. *Sovereign Wealth Funds and Long-Term Investing*. New York: Columbia University Press.

Borowiak, Malgorzata, Wallis Nahaboo, Martin Reynders, Katharina Nekolla, Pierre Jalinot, Jens Hasserodt, Markus Rehberg, et al. 2015. "Photoswitchable Inhibitors of Microtubules Dynamics Optimally Control Mitosis and Cell Death." *Cell* 162: 403–11.

Boulding, Kenneth E. 1966. "The Economics of the Coming Spaceship Earth." In *Environmental Quality in a Growing Economy*, ed. Kenneth E. Boulding, 3–14. Baltimore, MD: Resources for the Future/Johns Hopkins University Press.

Boykoff, Maxwell T., and Jules M. Boykoff. 2004. "Balance as Bias: Global Warming and the US Prestige Press." *Global Environmental Change* 14 (2): 125–36.

Brewster, David. 1855. *Memoirs of the Life, Writings, and Discoveries of Sir Isaac Newton*, vol. 2. Edinburgh: Thomas Constable.

Breyer, Stephen. 1993. *Breaking the Vicious Circle: Toward Effective Risk Regulation*. Cambridge, MA: Harvard University Press.

Brienen, Roel J. W., Oliver Lawrence Phillips, Ted R. Feldpausch, Manuel Gloor, T. R. Baker, Jonathan Lloyd, Gabriela Lopez-Gonzalez, et al. 2015. "Long-Term Decline of the Amazon Carbon Sink." *Nature* 519 (7543): 344–48.

Bristow, Charlie S., Karen A. Hudson-Edwards, and Adrian Chappell. 2010. "Fertilizing the Amazon and Equatorial Atlantic with West African Dust." *Geophysical Research Letters* 37. doi: 10.1029/2010GL043486.

Broothaerts, Wim, Heidi J. Mitchell, Brian Weir, Sarah Kaines, Leon M. A. Smith, Wei Yang, Jorge E. Mayer, Carolina Roa-Rodríguez, and Richard A. Jefferson. 2005. "Gene Transfer to Plants by Diverse Species of Bacteria." *Nature* 433: 629–31.

Buonocore, Jonathan J., Kathleen F. Lambert, Dallas Burtraw, Samantha Sekar, and Charles T. Driscoll. 2016. "An Analysis of Costs and Health Co-benefits for a U.S. Power Plant Carbon Standard." *PLoS ONE* 11 (6). doi: 10.1371/journal.pone.0156308.

Burke, Marshall, Solomon M. Hsiang, and Edward Miguel. 2015a. "Climate and Conflict." *Annual Review of Economics* 7: 577–617.

——. 2015b. "Global Non-linear Effects of Temperature on Economic Production." *Nature* 527: 235–39.

Byrd-Hagel Resolution, S. Res. 98, 105th Cong, 1st Sess (1997).

Caporaso, James. 1992. "International Relations Theory and Multilateralism: The Search for Foundations." *International Organization* 46 (3): 600–601.

Carbon Tracker Initiative. 2012. "Unburnable Carbon—Are the World's Financial Markets Carrying a Carbon Bubble." https://www.carbontracker.org/wp-content/uploads/2014/09/Unburnable-Carbon-Full-rev2-1.pdf.

Cardinale, Bradley J., J. Emmett Duffy, Andrew Gonzalez, David U. Hopper, Charles Perring, Patrick Venail, Anita Narwani, et al. 2012. "Biodiversity Loss and Its Impact on Humanity." *Nature* 486: 59–67.

Carpenter, Angela. 2012. "The Role of the European Union as a Global Player in Environmental Governance." *Journal of Contemporary European Research* 8 (2): 163.

Carson, Rachel. 1962. *Silent Spring.* New York: Houghton Mifflin.

Ceballos, Gerardo, Paul R. Ehrlich, Anthony D. Barnosky, Andrés García, Robert M. Pringle, and Todd M. Palmer. 2015. "Accelerated Modern Man-Induced Species Losses: Entering the Sixth Mass Extinction." *Science Advances* 1 (5): e1400253.

Chalfie, Martin, Yuan Tu, Ghia Euskirchen, William W. Ward, and Douglas C. Prasher. 1994. "Green Fluorescent Protein as a Marker to Gene Expression." *Science* 263: 802–5.

Charpentier, Emmanuelle, and Jennifer A. Doudna. 2013. "Biotechnology: Rewriting a Genome." *Nature* 495: 50–51.

China, People's Republic of. 2004. *Initial National Communication on Climate Change*. Beijing.

Chinese Agricultural University. 2009. *Fertilizer Application Guidelines for Main Crops of China*. Beijing: Chinese Agricultural University.

CNA Military Advisory Board. 2014. *National Security and the Accelerating Risks of Climate Change*. Alexandria, VA: CNA Corporation.

Coady, David, Ian Parry, Louis Sears, and Baoping Shang. 2015. "How Large Are Global Energy Subsidies?" IMF Working Paper WP/15/105. Washington, DC: International Monetary Fund.

Colombier, Michel, Denis Loyer, Laurence Tubiana, and Isabelle Biagiotti. 2011. "Une Négociation Climatique Plus Résiliente Qu'on ne le Craignait." In *Regards sur la Terre,* ed. Pierre Jacquet, Rajendra K. Pachauri, and Laurence Tubiana, 89–93. Paris: Armand Colin.

Commission on Intellectual Property Rights. 2002. *Integrating Intellectual Property Rights and Development Policy*. London: Department for International Development. http://www.iprcommission.org/papers/pdfs/final_report/CIPRfullfinal.pdf.

Committee on Himalayan Glaciers, Hydrology, Climate Change, and Implications for Water Security. 2012. *Himalayan Glaciers: Climate Change, Water Resources, and Water Security.* Washington, DC: National Academy of Sciences.

Crimmins, A., J. Balbus, J. L. Gamble, C. B. Beard, J. E. Bell, D. Dodgen, R. J. Eisen, et al., eds. 2016. *The Impacts of Climate Change on Human Health in the United States: A Scientific Assessment.* Washington DC: U.S. Global Change Research Program.

Crooks, Ed. 2013. "Saudis Welcome U.S. Shale Boom." *Financial Times,* May 13. https://www.ft.com/content/f84ebdeo-bbd4-11e2-82df-00144feab7de.

Daly, Herman E., ed. 1973. *Toward a Steady State Economy.* New York: W. H. Freeman.

Dasgupta, Partha. 2007. *Economics: A Very Short Introduction.* Oxford, UK: Oxford University Press.

Davenport, Coral. 2014. "As Oysters Die, Climate Policy Goes on the Stump." *New York Times,* August 3. https://www.nytimes.com/2014/08/04/us/as-oysters-die-climate-policy-goes-on-stump.html?_r=0.

Davies, Tim K., Chris C. Mees, and E. J. Milner-Gulland. 2016. "The Past, Present and Future Use of Drifting FADs in the Indian Ocean," *Marine Policy* 45: 163–70.

Davis, Aaron P., Tadesse Woldemariam Gole, Susana Baena, and Justin Moat. 2012. "The Impact of Climate Change on Indigenous Arabica Coffee (*Coffea arabica*): Predicting Future Trends and Identifying Priorities." *PLoS ONE* 7 (11): 1371–92.

Davis-Blake, Alison. 2015. Dean's Column in the Executive Education Magazine. *Financial Times,* May 17.

Davison, Nicola, and John Burn-Murdoch. 2016. "Invasive Species: The Battle to Beat the Bugs." *Financial Times,* May 16.

DeConto, Robert M., and David Pollard. 2016. "Contribution of Antarctica to Past and Future Sea-Level Rise." *Nature* 531: 591–97.

Descola, Philippe. 2013. *Beyond Nature and Culture.* Chicago: University of Chicago Press.

Diamanti-Kandarakis, Evanthia, Jean-Pierre Bourguignon, Linda C. Giudice, Russ Hauser, Gail S. Prins, Ana M. Soto, R. Thomas Zoeller, and Andrea C. Gore. 2009. "Endocrine-Disrupting Chemicals: An Endocrine Society Scientific Statement." *Endocrine Reviews* 30 (4): 293–342.

Dietrich, R., Sonja von Aulock, Hans Marquardt, Bas Blaauboer, Wolfgang Dekant, Jan Hengstler, James Kehrer, et al. 2013. "Scientifically Unfounded Precaution Drives European Commission's Recommendations on EDC Regulation, While Defying Sense, Well-Established Science and Risk Assessment Principles." *ALTEX* 30: 381–85.

Dinan, Terry. 2012. "Offsetting a Carbon Tax's Costs on Low-Income Households." CBO Working Paper 2012-16. Washington, DC: Congressional Budget Office.

Dolan, Ed. 2013. "A Carbon Tax May Curb the Rise in Natural Gas Flaring." OilPrice.com. http://oilprice.com/Energy/Natural-Gas/A-Carbon-Tax-may-Curb-the-Rise-in-Natural-Gas-Flaring.html.

Downs, Anthony. 1957. *An Economic Theory of Democracy*. New York: Harper.

Drake, Stillman. 1957. *Discoveries and Opinions of Galileo*. New York: Anchor.

Drijfhout, Sybren, Sebastian Bathiany, Claudie Beaulieu, Victor Brovkin, Martin Claussen, Chris Huntingford, Marten Scheffer, Giovanni Sgubin, and Didier Swingedouw. 2015. "Catalogue of Abrupt Shifts in Intergovernmental Panel on Climate Change Climate Models." *Proceedings of the National Academy of Sciences* 112 (43): E5777–E5786.

Ebbersmeyer, Curtis, and Eric Scigliano. 2009. *Flotsametrics and the Floating World: How One Man's Obsession with Runaway Sneakers and Rubber Ducks Revolutionized Ocean Science.* London: Collins.

Eck, Diana L. 1982. *Banāres: City of Light*. New York: Knopf.

——. 2013. *India: A Sacred Geography.* New York: Harmony Books.

Economist, The. 2006. "Clean Water Is a Right." International, November 9.

Economist, The. 2010. "The World's Lungs." Leaders, September 23.

Economy, Elizabeth C. 2004. *The River Runs Black: The Environmental Challenge to China's Future*. New York: Cornell University Press.

——. 2007. "The Great Leap Backward?" *Foreign Affairs* (September/October). https://www.foreignaffairs.com/articles/asia/2007-09-01/great-leap-backward.

Edwards, Guy, Isabell Cavalier Adarve, Maria Camila Bustos, and J. Tmmons Roberts. 2017. "Small Group, Big Impact: How AILAC Helped Shape the Paris Agreement." *Climate Policy* 17(1): 71–85.

Edwards, Guy, and Timmons Roberts. 2012. "A New Latin American Climate Negotiating Group : The Greenest Shoots in the Doha Desert." Brookings US Politics and government (blog), December 12. https://www.brookings.edu/blog/up-front/2012/12/12/a-new-latin-american-climate-negotiating-group-the-greenest-shoots-in-the-doha-desert/.

Edwards, Paul N. 2010. *A Vast Machine: Computer Models, Climate Data, and the Politics of Global Warming*. Boston: MIT Press.

Einstein, Albert. 1933. "On the Method of Theoretical Physics." Herbert Spencer Lecture, Oxford University, Oxford, UK, June 10.

ELD Initiative (Economics of Land Degradation) and UNEP (UN Environment Programme). 2015. *The Economics of Land Degradation in Africa: Benefits of Action Outweigh the Costs*. http://www.eld-initiative.org.

Energy Modeling Forum (EMF). 2013. "Changing the Game? Emissions and Market Implications of New Natural Gas Supplies." *EMF Report* 26 (1).

England, Matthew H., Alexander Sen Gupta, and Andrew J. Pitman. 2009. "Constraining Future Greenhouse Gas Emissions by a Cumulative Target." *Proceedings of the National Academy of Sciences* 106 (39): 16539–40.

Espagne, Etienne, Michel Aglietta, and Baptiste Perrissin Fabert. 2015. *A Proposal to Finance Low Carbon Investment in Europe*. Paris: Commissariat Général à la Stratégie et à la Prospective.

Etner, Johanna, Meglena Jeleva and Jean-Marie Tallon. 2012. "Decision Theory under Ambiguity." *Journal of Economic Surveys* 26(2): 234–70.

European Academies Science Advisory Council. 2015. "Ecosystem Services, Agriculture and Neonicotinoids." EASAC Policy Report 26. Halle, Germany: German National Academy of Sciences.

European Commission. 2016. "Commission Adopts First EU List of Invasive Alien Species, an Important Step Towards Halting Biodiversity Loss." http://ec.europa.eu/environment/pdf/13_07_2016_news_en.pdf

European Environment Agency. 2002. *Late Lessons from Early Warnings: The Precautionary Principle 1896–2000* (Environmental Issue Report 22). Copenhagen, Denmark: European Environment Agency.

European Parliament. 2002. *Report on the Proposal for a Council Decision Concerning the Conclusion, on Behalf of the European Community, of the Kyoto Protocol to the United Nations Framework Convention on Climate Change and the Joint Fulfilment of Commitments Thereunder*, A5-0025/2002. http://eur-lex.europa.eu/legal-content/EN/TXT/?uri=celex:52014PC0290 FINAL.

Falkner, Robert. 2010. "Getting a Deal on Climate Change Obama's Flexible Multilateralism." In *Obama Nation? U.S. Foreign Policy One Year On*, ed. Nicholas Kitchen, 37–41. London: LSE IDEAS Special Report.

Falkner, Robert, Hannes Stephan, and John Vogler. 2010. "International Climate Policy after Copenhagen: Towards a 'Building Blocks' Approach." *Global Policy* 1 (3): 252–62.

Fargione, Joseph, Jason Hill, David Tilman, Stephen Polasky, and Peter Hawthrone. 2008. "Land Clearing and the Biofuel Carbon Debt." *Science* 319 (5867): 1235–38.

Fermi, Laura, and Gilberto Bernardini. 2003. *Galileo and the Scientific Revolution.* New York: Dover.

Fisher, Matthew C., Daniel A. Henk, Cheryl J. Briggs, John S. Brownstein, Lawrence C. Madoff, Sarah L. McCraw, and Sarah J. Gurr. 2012. "Emerging Fungal Threats to Animal, Plant, and Ecosystem Health." *Nature* 484: 186–94.

Foley, Jonathan A., Ruth DeFries, Gregory P. Asner, Carol Bradford, Gordon Bonan, Stephen R. Carpenter, F. Stuart Chapin, et al. 2005. "Global Consequences of Land Use." *Science* 309 (5734): 570–74.

Fraas, Lewis M. 2014. *Low-Cost Solar Electric Power.* Heidelberg, Germany: Springer.

Freeman, Natalie M., and Nicole S. Lovenduski. 2015. "Decreased Calcification in the Southern Ocean over the Satellite Record." *Geophysical Research Letters* 42: 1834–40.

Freyman, Monika, and Ryan Salmon. 2013. "Hydraulic Fracturing and Water Stress: Growing Competitive Pressures for Water." Ceres Research Paper. Boston: Ceres.

Gao, Caixia, and Qui, Jin-Long. 2014. "Gene Edits Boost Wheat Defenses." *Nature* 511: 386–87.

Gastil, John, and Peter Levine, eds. 2005. *The Deliberative Democracy Handbook: Strategies for Effective Civic Engagement in the Twenty-First Century*. San Francisco: Jossey-Bass.

Gattuso, Jean-Pierre, Alexandre Magnan, R. Billé, William Cheung, Ella Howes, Fortunat Joos, Denis Allemand, et al. 2015. "Contrasting Futures for Ocean and Society from Different Anthropogenic Emissions Scenarios." *Science* 349 (6243): 1–10.

Geden, Oliver, and Stefan Schäfer. 2016. "'Negative Emissions': A Challenge for Climate Policy." SWP Comments 53. Berlin: Stiftung Wissenschaft und Politik.

Georgescu-Roegen, Nicolas. 1971. *The Entropy Law and the Economic Process*. Cambridge, MA: Harvard University Press.

Geyer, R., J. Jambeck, and K. L. Law. 2017. "Production, Use and Fate of All Plastics Ever Made." *Science Advances* 3 (7): e1700782. http://advances.sciencemag.org/content/3/7/e1700782.

Gilbert, Natasha. 2012. "African Agriculture: Dirt Poor." *Nature* 483 (7391): 525–27.

Gillis, Justin. 2010. "As Glaciers Melt, Science Seeks Data on Rising Seas." *New York Times,* November 13. http://www.nytimes.com/2010/11/14/science/earth/14ice.html.

Gittman, Rachel K., Charles H. Peterson, Carolyn A. Currin, F. Joel Fodrie, Michael F. Piehler, and John F. Bruno. 2016. "Living Shorelines Can Enhance the Nursery Role of Threatened Estuarine Habitats." *Ecological Applications* 26 (1): 249–63.

Global Alliance for Clean Cookstoves. 2015. "Five Years of Impact: 2010–2015." CleanCookstoves.org. http://cleancookstoves.org/resources/reports/fiveyears.html.

Global Coral Reef Monitoring Network, International Union for Conservation of Nature, and UN Environment Programme. 2012.

Glover, Jerry D., John P. Reganold, and Cindy M. Cox. 2012. "Agriculture: Plant Perennials to Save Africa's Soils." *Nature* 489: 359–61.

Goodstein, David. 2010. *On Fact and Fraud: Cautionary Tales from the Front Lines of Science*. Princeton, NJ: Princeton University Press.

Goulson, Daniel. 2014. *A Buzz in the Meadow: The Natural History of a French Farm*. New York: Picador.

Grandjean, Philippe. 2015. *Only One Chance: How Environmental Pollution Impairs Brain Development—and How to Protect the Brains of the Next Generation*. Environmental Ethics and Science Policy Series. New York: Oxford University Press.

Gray, Erin, Peter G. Veit, Juan Carlos Altamirano, Helen Ding, Piotr Rozwalka, Ivan Zuniga, Matthew Witkin, et al. 2015. "The Economic Costs and Benefits of Securing Community Forest Tenure: Evidence from Brazil and Guatemala." World Resources Institute Working Paper. Washington, DC: World Resources Institute.

Green, Duncan. 2016. *How Change Happens*, Oxford: Oxford University Press.

Green Growth Action Alliance. 2013. *The Green Investment Report: The Ways and Means to Unlock Private Finance for Green Growth.* Geneva: World Economic Forum.

Greer, Amy, Victoria Ng, and David Fisman. 2008. "Climate Change and Infectious Diseases in North America: The Road Ahead." *Canadian Medical Association Journal* 178 (6): 715–22.

Groen, Lisanne, Arne Niemann, and Sebastian Oberthür. 2012. "The EU as a Global Leader? The Copenhagen and Cancún UN Climate Change Negotiations." *Journal of Contemporary European Research* 8 (2): 173–91.

Grogan, James, Christopher Free, Gustavo Pinelo Morales, Andrea Johnson, Rubí Alegria, and Benjamin Hodgdon. 2015. "Sustaining the Harvest: Assessment of the Conservation Status of Big-Leaf Mahogany, Spanish Cedar, and Three Lesser-Known Timber Species Populations in the Forestry Concessions of the Maya Biosphere Reserve, Petén, Guatemala." *Community Forest Case Studies* 5 (10). http://www.rainforest-alliance.org/sites/default/files/2016-08/sustaining-the-harvest.pdf.

Grubb, Michael, and Joyeeta Gupta, eds. 2000. *Climate Change and European Leadership: A Sustainable Role for Europe?* Dordrecht, Netherlands: Kluwer Academic.

Guo, Jingheng, Xuejun Liu, Yong Zhang, Jianlin Shen, Xin-Wei Han, Weifeng Zhang, P. Christie, et al. 2010. "Significant Acidification in Major Chinese Croplands." *Science* 327 (5968): 1008–10.

Guo, Wanqin, Junli Xu, Shiyin Liu, Donghui Shangguan, Lizong Wu, Xiaojun Yao, Jingdong Zhao, et al. 2014. *The Second Glacier Inventory Dataset of China (Version 1.0): Cold and Arid Regions.* Lanzhou, China: Science Data Center.

Haas, Peter M. 1989. "Do Regimes Matter? Epistemic Communities and Mediterranean Pollution Control." *International Organization* 43 (3): 377–403.

——. 1992. "Epistemic Communities and International Policy Coordination." *International Organization* 46 (1): 1–35.

Halpern, Benjamin S., and Robert R. Warner. 2002. "Marine Reserves Have Rapid and Lasting Effects." *Ecology Letters* 5 (3): 361–66.

Hansen, James, Makiko Sato, Pushker Kharecha, David Beerling, Valerie Masson-Delmotte, Mark Pagani, Maureen Raymo, Dana L. Royer, and James C. Zachos. 2008. "Target Atmospheric CO_2: Where Should Humanity Aim?" *Open Atmospheric Science Journal* 2: 217–31.

Hansen, James, Makiko Sato, and Reto Ruedy. 2012. "Perception of Climate Change." *Proceedings of the National Academy of Science* 109 (37): E2415–23.

Hardin, Garett. 1968. "The Tragedy of the Commons." *Science* 162 (3859): 1243–48.

Harhoff, Dietmar, Pierre Régibeau, Katharine Rockett, Monika Schnitzer, and Bruno Jullien. 2001. "Some Simple Economics of GM Food." *Economic Policy* 16 (33): 263–99.

Harremoës, Poul, David Gee, Malcolm MacGarvin, Andy Stirling, Jane Keys, Brian Wynne, and Sofia Guedes Vaz, eds. 2001. "Late Lessons from Early Warnings: The Precautionary Principle 1896–2000." Environmental Issue Report 22. Copenhagen: European Environment Agency.

Hawkes, L. A., A. C. Broderick, M. H. Godfrey, and B. J. Godley. 2007. "Investigating the Potential Impacts of Climate Change on a Marine Turtle Population." *Global Change Biology* 13 (5): 923–32.

Heal, Geoffrey M. 2016. *Endangered Economies: How the Neglect of Nature Threatens Our Prosperity*, New York: Columbia University Press.

——. 2000. *Nature and the Marketplace: Capturing the Value of Ecosystem Services.* Washington, DC: Island Press.

Heal, Geoffrey, Antony Millner, and Simon Dietz. 2010. "Ambiguity and Climate Policy." Columbia Business School and NBER Working Paper 16954. Cambridge MA: National Bureau of Economic Research.

Henry, Claude. 2010. "Decision-Making Under Scientific, Political and Economic Uncertainty." In *The Beijer Institute at the Royal Swedish Academy of Sciences—Bringing Ecologists and Economists Together: The Askö Meetings and Papers.* Munich: Springer Verlag.

Henry, Claude, and Joseph E. Stiglitz. 2010. "Intellectual Property, Dissemination of Innovation and Sustainable Development." *Global Policy* 1 (3): 237–51.

Henry, Emeric. 2010. "Runner-Up Patents: Is Monopoly Inevitable?" *Scandinavian Journal of Economics* 112 (2): 417–40.

Henry, Terrence, and Kate Galbraith. 2013. "As Fracking Proliferates, So Do Waste Water Wells." *New York Times,* March 28, A21.

Hermitte, Sam Marie, and Robert E. Mace. 2012. "The Grass Is Always Greener . . . Outdoor Residential Water Use in Texas." Technical Note 12–01. Austin, TX: Texas Water Development Board.

Hillel, Daniel. 2006. *The Natural History of the Bible: An Environmental Exploration of the Hebrew Scriptures.* New York: Columbia University Press.

——. 2008. *Soil in the Environment: Crucible of Terrestrial Life.* Burlington, MA: Academic Press.

——. 2012. "Statement of Achievement of the Laureate." https://www.worldfoodprize.org/en/laureates/2010_2015_laureates/2012_hillel/#StatementAchievement.

Hirschfeld Davis, Julie, Mark Landler, and Coral Davenport. 2016. "Obama on Climate Change: The Trends Are 'Terrifying.'" *New York Times*, September 8. https://www.nytimes.com/2016/09/08/us/politics/obama-climate-change.html?_r=0.

Hochstetler, Kathryn Ann. 2012. "The G-77, BASIC, and Global Climate Governance: A New Era in Multilateral Environmental Negotiations." *Revista Brasileira de Política Internacional* 55 (special edition): 53–69.

Hoggan, James. 2009. *Climate Cover-Up: The Crusade to Deny Global Warming.* Vancouver: Greystone.

Hopwood, Jennifer, Scott Hoffman Black, Mace Vaughan, and Eric Lee-Mäder. 2013. *Beyond the Birds and the Bees: Effects of Neonicotinoid Insecticides on Agriculturally Important Beneficial Invertebrates.* Portland, OR: Xerces Society for Invertebrate Conservation.

Horel, Stéphane. 2015. *Intoxication: Perturbateurs Endocriniens, Lobbyistes, et Eurocrates: Une Bataille d'Influence Contre la Santé.* Paris: La Découverte.

Hsiang, Solomon, Robert Kopp, D. J. Rasmussen, Michael Mastrandrea, Amir Jina, James Rising, Robert Muir-Wood, et al. 2014. "American Climate Prospectus: Economic Risks in the United States." Goldman School of Public Policy Working Paper Series (prepared as input to the Rhodium Group Risky Business Project). Berkeley: University of California.

Huang, Xingxu, et al. 2014. "CRISPR Makes Modified Monkeys." *Nature* 506: 8.

Ilsted, Ulrik, Aida Bargués Tobella, Bazié Hugues Roméo, Jules Bayala, Verbeeten Elke, Gert Nyberg, Josias Sanou, et al. 2016. "Intermediate Tree Cover Can Maximize Groundwater Recharge in the Seasonally Dry Tropics." *Scientific Reports* 6: 21930.

Im, Eun-Soon, Jeremy S. Pal, and Elfatih A. B. Eltahir. 2017. "Deadly Heat Waves Projected in the Densely Populated Agricultural Regions of South Asia." *Science Advances* 3 (8): e1603322. http://advances.sciencemag.org/content/3/8/e1603322.

Innis, Lorna, and Alan Simcock, eds. 2016. *The First Global Integrated Marine Assessment.* New York: United Nations General Assembly.

Inter-American Development Bank. 2012. *Climate Change Action Plan 2012–15*, Washington, DC: IADB.

Intergovernmental Panel on Climate Change. 2014. *Fifth Assessment Report.* Geneva: World Health Organization and UN Environment Program.

Intergovernmental Science-Policy Platform on Biodiversity and Ecosystem Services. 2016. *The Assessment Report on Pollinators, Pollination, and Food Production.* Bonn, Germany: IPBES Secretariat.

International Centre of Insect Physiology and Ecology. 2011. *Planting for Prosperity: Push-Pull: A Model for Africa's Green Revolution.* Duduville, Nairobi: ICIPE.

International Energy Agency. 2006. "Energy for Cooking in Developing Countries." In *World Energy Outlook 2006,* chap. 15. Paris: Organisation for Economic Co-operation and Development.

——. 2012. *World Energy Outlook*, Executive Summary (English). http://www.iea.org/publications/freepublications/publication/English.pdf.

——. 2015a. *2015 Energy Efficiency Market Report.* Paris: Organisation for Economic Co-operation and Development.

——. 2015b. *World Energy Outlook 2015.* Paris: Organisation for Economic Co-operation and Development.

——. 2016. *World Energy Outlook 2016.* Paris: Organisation for Economic Co-operation and Development.

International Energy Agency, Organisation for Economic Co-operation and Development, and World Bank. 2010. "The Scope of Fossil-Fuel Subsidies in 2009 and a Roadmap for Phasing Out Fossil-Fuel Subsidies." Joint report prepared for the G-20 Summit, Seoul, Korea, November 11–12.

International Institute for Sustainable Development. 2001. *Earth Negotiations Bulletin*, 1.

Islam, A. K. M. Sadrul, and Mohammad Ahiduzzaman. 2013. "Green Electricity from Rice Husk: A Model for Bangladesh." In *Thermal Power Plants—Advanced Applications*, ed. Mohammad Rasul. Rijeka, Croatia: InTech Open Access.

Jaffe, Adam B., and Josh Lerner. 2004. *Innovation and Its Discontents: How Our Broken Patent System Is Endangering Innovation and Progress, and What to Do About It.* Princeton, NJ: Princeton University Press.

Jansson, Peter, Regine Hock, and Thomas Schneider. 2003. "The Concept of Glacier Storage: A Review." *Journal of Hydrology* 282 (1–4): 116–29.

Jasanoff, Sheila. 2005. *Designs on Nature: Science and Democracy in Europe and the United States.* Princeton, NJ: Princeton University Press.

Jasanoff, Sheila, and Marybeth Long Martello, eds. 2004. *Earthly Politics. Local and Global in Environmental Governance* (Politics, Science, and the Environment Series). Cambridge, MA: MIT Press.

Jing, Li. 2010. "Safeguarding Environment a Priority." *China Daily*, November 29. http://www.chinadaily.com.cn/china/2010-11/29/content_11620813.htm.

Johnson, Dexter. 2015. "Supercapacitors Take Huge Leap in Performance." *IEEE Spectrum,* May 28.

Jones, Bryan, Brian C. O'Neill, Larry McDaniel, Seth McGinnis, Linda O. Mearns, and Claudia Tebaldi. 2015. "Future Population Exposure to US Heat Extremes." *Nature Climate Change* 5: 652–55.

Jones, Nicola. 2006. "Climate Change Blamed for India's Monsoon Misery." *Nature News*, November 14.

——. 2009. "Climate Crunch: Sucking It Up." *Nature* 458: 1094–97.

——. 2011. "Climate Change Curbs Crops: Warming Already Lowered Yields of Wheat and Corn." *Nature,* May 5.

Joskow, Paul L. 2013. "Natural Gas: From Shortages to Abundance in the United States." *American Economic Review* 103 (3): 338–43.

Ju, X.-T. et al. 2009. "Reducing Environmental Risk by Improving N Management in Intensive Chinese Agricultural Systems." *PNAS* 106: 3041–46.

Karp, Daniel S., Chase D. Mendenhall, Randi Figueroa Sandí, Nicolas Chaumont, Paul R. Ehrlich, Elizabeth A. Hadly, and Gretchen C. Daily. 2013. "Forest Bolsters Bird Abundance, Pest Control, and Coffee Yield." *Ecology Letters* 16 (11): 1339–47.

Kaul, Inge, Isabelle Grunberg, and Marc Stern, eds. 1999. *Global Public Goods: International Cooperation in the 21st Century*. Oxford: Oxford University Press.

Kearns, Cristin E., Stanton A. Glantz, and Laura A. Schmidt. 2015. "Sugar Industry Influence on the Scientific Agenda of the National Institute of Dental Research's National Caries Program: A Historical Analysis of Internal Documents." *PLoS Medicine* 12 (3): E1001798.

Kearns, Cristin E., Laura A. Schmidt, and Stanton A. Glantz. 2016. "Sugar Industry and Coronary Heart Disease Research: A Historical Analysis of Internal Industry Documents." *JAMA Internal Medicine* 176: 1680–85.

Kegley, Charles W. Jr., and Gregory Raymond. 1994. *A Multipolar Peace? Great Powers Politics in the Twenty-First Century*. New York: St Martin's.

Kelemen, Peter B., and Jürg Matter. 2008. "In Situ Carbonation of Peridotite for CO_2 Storage." *Proceedings of the National Academy of Sciences* 105 (45): 17295–300.

Kelemen, Peter B., Jürg Matter, Elisabeth E. Streit, John F. Rudge, William B. Curry, Jerzy Blusztajn. 2011. "Rates and Mechanisms of Mineral Carbonation in Peridotite: Natural Processes and Recipes for Enhanced in Situ CO_2 Capture and Storage." *Annual Review of Earth and Planetary Sciences* 39: 545–76.

Kennedy, Kevin. 2013. "California's Cap-and-Trade Program Makes Encouraging Headway" (blog). http://www.wri.org/blog/2013/08/california%E2%80%99s-cap-and-trade-program-makes-encouraging-headway.

Keohane, Robert O. 1990. "Multilateralism: An Agenda for Research." *International Journal* 45: 731.

Keohane, Robert O., and David G. Victor. 2010. "The Regime Complex for Climate Change." Discussion Paper 2010-33. Cambridge, MA: Harvard Project on International Climate Agreements.

Keynes, John Maynard. 1921/2016. *A Treatise on Probability*. London: Macmillan.

Kim, Joo Gon, Byungrak Son, Santanu Mukherjee, Nicholas Schuppert, Alex Bates, Osung Kwon, Moon Jong Choi, Hyun Yeoi Chung, and Sam Park. 2015. "A Review of Lithium and Non-Lithium Based Solid-State Batteries." *Journal of Power Sciences* 282: 299–322.

King, H. 2013. "Natural Gas Flaring in North Dakota." Geology.com.

Klein, Naomi. 2014. *This Changes Everything: Capitalism Versus the Climate*. New York: Simon & Schuster.

Klibanoff, Peter, Massimo Marinacci, and Sujoy Mukerji. 2005. "A Smooth Model of Decision Making Under Ambiguity." *Econometrica* 73 (6): 1849–92.

Knutti, Reto, Joeri Rogelj, Jan Sedlacek, and Erich M. Fischer. 2015. "A Scientific Critique of the Two-Degree Climate Change Target." *Nature Geoscience* 9: 13–18.

Kolbert, Elizabeth. 2006. "The Climate of Man." *New Yorker*.

——. 2014. *The Sixth Extinction: An Unnatural History*. London: Bloomsbury.

Kort, E., et al. 2012. "Atmospheric Observations of Arctic Ocean Methane Emissions up to 82° North." *Nature Geoscience* 5: 318–21.

Kortenkamp, Andreas, Olwenn Martin, Michael Faust, Richard Evans, Rebecca McKinlay, Frances Orton, and Erika Rosivatz. 2011. "State of the Art Assessment of Endocrine Disrupters." Final Report to the European Commission, Environment Directorate-General.

Kroto, Harold. 1997. "Symmetry, Space, Stars, and C_{60} (Nobel Lecture)." *Angewandtë Chemie* 36 (15): 1578–93.

Kuhn, Thomas S. 1962. *The Structure of Scientific Revolutions*. Chicago: University of Chicago Press.

Lackner, Klaus S., Sarah Brennan, Jürg M. Matter, A.-H. Alissa Park, Allen Wright, and Bob van der Zwaan. 2012. "The Urgency of the Development of CO_2 Capture from Ambient Air." *Proceedings of the National Academy of Sciences* 109 (33): 13156–162.

Lancet. 2012. "A Comparative Risk Assessment of Disease and Injury Attributable to 67 Risk Factors."

Leaton, James. 2012. *Unburnable Carbon: Are the World's Financial Markets Carrying a Carbon Bubble?* London: Carbon Tracker Initiative. https://www.carbontracker.org/wp-content/uploads/2014/09/Unburnable-Carbon-Full-rev2-1.pdf.

Lerin, François, and Laurence Tubiana. 2005. "Questions Autour de l'Agenda Environnemental International." *Revue Internationale et Stratégique* 4 (60): 75–84.

Leslie, Jacques. 2014. "Is Canada Tarring Itself?" (op-ed). *New York Times*, March 30, A21.

Lévêque, François. 2014. *The Economics and Uncertainties of Nuclear Power*. Cambridge, UK: Cambridge University Press

Levi, Michael. 2013. "Climate Consequences of Natural Gas as a Bridge Fuel." *Climatic Change* 118 (3): 609–23.

Levy, David L., and Peter J. Newell, eds. 2005. *The Business of Global Environmental Governance*. Cambridge, MA: MIT Press.

Lin, Kaixiang, Qing Chen, Michael R. Gerhardt, Liuchuan Tong, Sang Bok Kim, Louise Eisenach, Alvaro W. Valle, et al. 2015. "Alkaline Quinone Flow Battery." *Science* 349 (6255): 1529–32.

Liu, Xuejun, Ying Zhang, Wenxuan Han, Aohan Tang, Jianlin Shen, Zhenling Cui, Peter Vitousek, et al. 2013. "Enhanced Nitrogen Deposition over China." *Nature* 494: 459–62.

Lobell, David. 2011. "Climate Change and Agricultural Adaptation." Paper presented at the Stanford Symposium Series on Global Food Policy and Food Security in the 21st Century, December 8, Stanford, CA.

Lomborg, Bjørn. 2001. *The Skeptical Environmentalist: Measuring the Real State of the World*. Cambridge, UK: Cambridge University Press.

Londo Jason, Nonnatus S. Bautista, Cynthia L. Sagers, E. Henry Lee, and Lidia S. Watrud. 2010. "Glyphosate Drift Promotes Change in Fitness and Transgene Gene Flow in Canola (*Brassica napus*) and Hybrids." *Annals of Botany* 106 (6): 957–65.

Lu, Yanhui, Kongming Wu, Yuying Jiang, Bing Xia, Ping Li, Hongqiang Feng, Kris Wyckhuys, and Yuyan Guo. 2010. "Mirid Bug Outbreaks in Multiple Crops Correlated with Wide-Scale Adoption of Bt Cotton in China." *Science* 328 (5982): 1151–54.

Lumban Gaul, Amy. 2015. "Surviving the Long Dry Season in Konawe Selatan with Improved Farming Systems." CGIAR Research Program on Forests, Trees and Agroforestry: Research Paper, November 6.

Malla, Sunil, and Govinda R. Timilsina. 2014. "Household Cooking Fuel Choice and Adoption of Improved Cookstoves in Developing Countries: A Review." Policy Research Working Paper 6903. Washington, DC: World Bank.

Malthus, Robert Thomas. 1836. *Principles of Political Economics*. London: Pickering.

Manzanares, F. Javier. 2017. "The Green Climate Fund–A Beacon for Climate Change Action", *Asian Journal for Sustainability and Social Responsibility* 12. https://doi.org/10.1186/s41180-016-0012-1.

Marchese, Anthony J., Timothy L. Vaughn, Daniel J. Zimmerle, David M. Martinez, Laurie L. Williams, Allen L. Robinson, Austin L. Mitchell, et al. 2015. "Methane Emissions from United States Natural Gas Gathering and Processing." *Environmental Science and Technology* 49 (17): 10718–27.

Martin, Rosa María Fernández. 2012. "The European Union and International Negotiations on Climate Change. A Limited Role to Play," *Journal of Contemporary European Research* 8 (2): 192–209.

Maskin, Eric S. 2008. "Mechanism Design: How to Implement Social Goals (Nobel Lecture)." *American Economic Review* 98 (3): 567–76.

Masood, Salman. 2015. "Starved for Energy, Pakistan Braces for a Water Crisis." *New York Times,* February 12. https://www.nytimes.com/2015/02/13/world/asia/pakistan-braces-for-major-water-shortages.html.

Matter, Jürg, Martin Stute, Sandra O. Snaebjörnsdottir, Eric H. Oelkers, Sigurdur R. Gislason, Edda S. Aradottir, Bergur Sigfusson, et al. 2016. "Rapid Carbon Mineralization for Permanent Disposal of Anthropogenic Carbon Dioxide Emissions." *Science* 352 (6291): 1312–14.

McPhee, John. 1974. *The Curve of Binding Energy: A Journey into the Awesome and Alarming World of Theodore B. Taylor.* New York: Farrar, Straus and Giroux.

Meadows, Dennis H., Donella L. Meadows, Jørgen Randers, and William W. Behrens III. 1972. *The Limits to Growth*. New York: Universe Books.

Medina, Jennifer. 2014. "In California, Climate Issues Moved to Fore by Governor." *New York Times,* May 19. https://www.nytimes.com/2014/05/20/us/politics/in-california-climate-issues-moved-to-fore-by-governor.html.

Meinshausen, Maite, Nicolai Meinshausen, William Hare, Sarah C. B. Raper, Katja Frieler, Reto Knutti, David J. Frame, and Myles R. Allen. 2009. "Greenhouse-Gas Emissions Targets for Limiting Global Warming to 2°C." *Nature* 458: 1158–62.

Merchant, Carolyn. 1980. *The Death of Nature: Women, Ecology, and the Scientific Revolution*, New York: HarperCollins.

Merges, Robert P., and Richard R. Nelson. 1990. "On the Complex Economics of Patent Scope." *Columbia Law Review* 90 (4): 839–916.

Metcalf, Gilbert E. 2009. "Market-Based Policy Options to Control U.S. Greenhouse Gas Emissions." *Journal of Economic Perspectives* 23 (2): 5–27.

Metcalf, Gilbert E., and David Weisbach. 2009. "The Design of a Carbon Tax." *Harvard Environmental Law Review* 33: 499–506.

Miller, Clark A., and Paul N. Edwards, eds. 2001. *Changing the Atmosphere: Expert Knowledge and Environmental Governance*. Cambridge, MA: MIT Press.

Millner, Antony, Simon Dietz, and Geoffrey Heal. 2013. "Scientific Ambiguity and Climate Policy." *Environmental and Resource Economics* 55(1): 21–46.

Milne, Janet E., and Mikael S. Andersen, eds. 2012. *Handbook of Research on Environmental Taxation*. Cheltenham: Edward Elgar.

Milton, John. 1644. *Areopagitica*. https://www.dartmouth.edu/~milton/reading_room/areopagitica/text.html

Mnif, Wissem, Aziza Ibn Hadj Hassine, Aicha Bouaziz, Aghleb Bartegi, Olivier Thomas, and Benoit Roig. 2011. "Effects of Endocrine Disruptor Pesticides: A Review." *International Journal of Environmental Resources and Public Health* 8 (6): 2265–2303.

Mokyr, Joel. 2004. *The Gifts of Athena: Historical Origins of the Knowledge Economy*. Princeton, NJ: Princeton University Press.

Molinos, Jorge García, Benjamin S. Halpern, David S. Schoeman, Christopher J. Brown, Wolfgang Kiessling, Pippa J. Moore, John M. Pandolfi, et al. 2016. "Climate Velocity and the Future Global Redistribution of Marine Biodiversity." *Nature Climate Change* 6: 83–88.

Montgomery, David R. 2010. "2010 Visions: Soil," *Nature* 463: 31–32.

Montgomery, David R. 2012. *Dirt: The Erosion of Civilizations*. Berkeley: University of California Press.

Morgan, Jennifer, and Kevin Kennedy. 2013. "First Take: Looking at President Obama's Climate Action Plan" (blog). http://www.wri.org/blog/2013/06/first-take-looking-president-obama%E2%80%99s-climate-action-plan.

Morgan, Jennifer, and Deborah Seligsohn. 2010. "What Cancun Means for China and the U.S." ChinaFAQs, December 15. http://www.chinafaqs.org/blog-posts/what-cancun-means-china-and-us.

Morrisette, Peter M. 1989. "The Evolution of Policy Responses to Stratospheric Ozone Depletion." *Natural Resources Journal* 29: 793–820.

Mortensen, David A., J. Franklin Egan, Bruce D. Maxwell, Matthew R. Ryan, and Richard G. Smith. 2012. "Navigating a Critical Juncture for Sustainable Weed Management." *BioScience* 62 (1): 75–84.

Moss, Michael. 2014. *Salt Sugar Fat: How the Food Giants Hooked Us*. New York: Random House.

Mouginot, J., E. Rignot, B. Scheuchl, I. Fenty, A. Khazendar, M. Morlighem, A. Buzzi, and J. Paden. 2015. "Fast Retreat of Zachariae Isstrøm, Northeast Greenland." *Science* 350 (6266): 1357–61.

Mowery, David C., Richard R. Nelson, and Ben R. Martin. 2010. "Technology Policy and Global Warming: Why New Policy Models Are Needed." *Research Policy* 39 (8): 1011–23.

Najam, Adil, Mihaela Papa, and Nadaa Taiyab. 2006. *Global Environmental Governance: A Reform Agenda*. Winnipeg, Manitoba, Canada: International Institute for Sustainable Development.

National Academies of Sciences, Engineering, and Medicine. 2016. *Attribution of Extreme Weather Events in the Context of Climate Change.* Washington, DC: National Academies Press. doi: 10.17226/21852.

National Research Council. 2008. *Severe Space Weather Events—Understanding Societal and Economic Impacts: A Workshop Report.* Washington, DC: National Academies Press.

——. 2015. *Climate Intervention: Carbon Dioxide Removal and Reliable Sequestration.* Washington, DC: National Academies Press.

Natural Resources Conservation Service (NRCS). 2017. "Insects and Pollinators." www.nrcs.usda.gov/wps/portal/nrcs/detailfull/national/plantsanimals/pollinate/?cid=nrcsdev11_000188

Nature. 2010. "2020 Visions." Opinions, January 7, 26–32. doi: 10.1038/463026a.

Nature. 2013. "365 Days: Nature's 10: Ten People Who Mattered This Year." December 18. https://www.nature.com/news/365-days-nature-s-10-1.14367.

Nayar, Anjali. 2012. "African Land Grabs Hinder Sustainable Development." *Nature* 481. http://www.nature.com/news/african-land-grabs-hinder-sustainable-development-1.9955.

Newbold, Tim, Lawrence N. Hudson, Andrew P. Arnell, Sara Contu, Adriana De Palma, Simon Ferrier, Samantha L. L. Hill, et al. 2016. "Has Land Use Pushed Terrestrial Biodiversity Beyond Planetary Boundary? A Global Assessment." *Science* 353 (6296): 288–91.

Newfarmer, Richard. 2001. *Global Economic Prospects and the Developing Countries 2002: Making Trade Work for the World's Poor.* Washington, DC: World Bank.

Niu Yuyu, Xingxu Huang, et al. 2014. "Generation of Gene-Modified Cynomolgus Monkey via Cas9/RNA Mediated Gene Targeting in One-Cell Embryos." *Cell* 156: 836–43.

North, Douglass. 1993. "Economic Performance Through Time" (Nobel Lecture). http://www.nobelprize.org/nobel_prizes/economic-sciences/laureates/1993/north-lecture.html.

Nye, Bill. 2016. *Unstoppable: Harnessing Science to Change the World.* New York: St. Martin's.

Nysveen, Per Magnus. 2016. "United States Now Holds More Oil Reserves Than Saudi Arabia" (press release). Oslo, Norway: Rystad Energy.

Olah, George A., G. K. Surya Prakash, and Alain Goeppert. 2011. "Anthropogenic Chemical Carbon Cycle for a Sustainable Future." *Journal of the American Chemical Society* 133 (33): 12881–98.

Olson, Mancur. 1965. *The Logic of Collective Action: Public Goods and the Theory of Groups.* Cambridge, MA: Harvard University Press.

O'Neill, Kate. 2007. "From Stockholm to Johannesburg and Beyond: The Evolving Meta-Regime for Global Environmental Governance." Paper presented at the 2007 Amsterdam Conference on the Human Dimensions of Global Environmental Change, May 24–26.

Oreskes, Naomi, and Erik M. Conway. 2010. *Merchants of Doubt: How a Handful of Scientists Obscured the Truth on Issues from Tobacco Smoke to Global Warming.* New York: Bloomsbury.

Organisation for Economic Co-operation and Development. 2010. "Annex A." In *Fiscality, Innovation, and Environment.* Paris: OECD Publishing.

——. 2016. *Agricultural Policy Monitoring and Evaluation 2016.* Paris: OECD Publishing.

Ostrom, Elinor. 1990. *Governing the Commons: The Evolution of Institutions for Collective Action.* New York: Cambridge University Press.

Overpeck, Jonathan T., and Jeremy L. Weiss. 2009. "Projections of Future Sea Level Becoming More Dire." *Proceedings of the National Academy of Sciences* 106 (51): 21461–62.

Oxfam International. 2009. *Bolivia: Climate Change, Poverty, and Adaptation.* La Paz, Bolivia: Oxfam International.

Painter, James. 2007. "Deglaciation in the Andean Region." Human Development Report 2007/55. New York: UN Development Programme.

Palumbi, Stephen R. 2001. "Humans as the World's Greatest Evolutionary Force." *Science* 293 (5536): 1786–90.

Park, Jacob, Ken Conca, and Matthias Finger, eds. 2008. *The Crisis of the Global Environmental Governance: Towards a New Political Economy of Sustainability.* Oxon, UK: Routledge.

Park Williams, A., Richard Seager, John T. Abatzoglou, Benjamin I. Cook, Jason E. Smerdon, and Edward R. Cook. 2015. "Contribution of Anthropogenic Warming to California Drought During 2012–2014." *Geophysical Research Letters* 42 (16): 6819–28.

Payne, Jonathan L., Andrew M. Bush, Noel A. Heim, Matthew L. Knope, and Douglas J. McCauley. 2016. "Ecological Selectivity of the Emerging Extinction in the Oceans." *Science* 353: 1284–86.

Pennisi, Elizabeth. 2009. "Systematics Researchers Want to Fend Off Patents." *Science* 325 (5941): 664.

Peplow, Mark. 2014. "Cellulosic Ethanol Fights for Life." *Nature* 507: 152–53.

Pereira, Henrique M., Paul W. Leadley, Vania Proenca, Rob Alkemade, Jörn P. W. Scharlemann, Juan F. Fernandez-Manjarrés, and Miguel B. Araújo, et al. 2010. "Scenarios for Global Biodiversity in the 21st Century." *Science* 330 (6010): 1496–1501.

Perry, Mike L. 2015. "Expanding the Chemical Space for Redox Flow Batteries." *Science* 349 (6255): 1452–53.

Pew Center on Global Climate Change. 2009. "Key Provisions American Recovery and Reinvestment Act: Energy, Transport, and Climate Research Spending in ARRA."

Philibert, Cédric. 2011. *Solar Energy Perspectives.* Paris: International Energy Agency.

Pollack, Henry N. 1997. *Uncertain Science . . . Uncertain World.* Cambridge, UK: Cambridge University Press.

Pope, C. Arden, and Douglas W. Dockery. 2013. "Air Pollution and Life Expectancy in China and Beyond." *Proceedings of the National Academy of Sciences of the United States of America* 110(32): 12861–62.

Popkin, Gabriel. 2015. "Breaking the Waves." *Science* 350 (6262): 756–59.

Potsdam Institute for Climate Impact Research and Climate Analytics. 2012. *Turn Down the Heat: Why a 4°C Warmer World Must Be Avoided.* Washington DC: World Bank.

Powell, Lindsey. 2003. *In Defense of Multilateralism*. New Haven, CT: Yale Center for Environmental Law and Policy.

Pozen, Robert C. Interview in *The New York Times*, November 16, 2009.

President's Science Advisory Committee (PSAC). 1963. "Pesticides Report." https://www.jfklibrary.org/Asset-Viewer/Archives/JFKPOF-087-003.aspx.

Pretty, Jules N., A. D. Noble, D. Bossio, J. Dixon, R. E. Hine, F.W. T. Penning de Vries, and J. I. L. Morrison. 2006. "Resource-Conserving Agriculture Increases Yields in Developing Countries." *Environmental Science and Technology* 40 (4): 1114–19.

Putnam, Robert D. 1988. "Diplomacy and Domestic Politics: The Logic of the Two-Level Games." *International Organization* 42 (3): 427–60.

Rainforest Action Network. "Protecting the Leuser Ecosystem—A Shared Responsibility." San Francisco: Rainforest Action Network. https://d3n8a8pro7vhmx.cloudfront.net/rainforestactionnetwork/pages/17068/attachments/original/1478466859/RAN_Protecting_The_Leuser_2016.pdf?1478466859.

Rambicur, Jean-François, and François Jaquenoud. 2013. "1,001 Fontaines pour Demain: Pour Une Nouvelle Économie de l'Eau Potable." *Le Journal de l'Ecole de Paris* 102: 25–31.

Ramstein, Céline. 2012. "Rio+20 Voluntary Commitments: Delivering Promises on Sustainable Development?" IDDRI Working Paper No. 23/12. Paris: Institut du Développement Durable et des Relations Internationales.

Rask, Mikko, Richard Worthington, and Minna Lammi, eds. 2012. *Citizen Participation in Global Environmental Governance*. Oxon, UK: Earthscan.

Raustiala, Kal. 2005. "Form and Substance in International Agreements." *American Journal of International Law*, 99 (3): 581–614.

Raworth, Kate. 2012. "A Safe and Just Space for Humanity: Can We Live Within the Doughnut?" Oxfam Discussion Paper. Oxford, UK: Oxfam International.

Reddy, Sai Bhaskar Nakka. 2012. *Understanding Stoves: For Environment and Humanity.* (Self-published with CreateSpace).

Reuters. 2014. "China Aims High for Carbon Market by 2020." *Sydney Morning Herald*, September 12. http://www.smh.com.au/environment/climate-change/china-aims-high-for-carbon-market-by-2020-20140911-10ftdp.html.

Richey, Alexandra S., Brian F. Thomas, Min-Hui Lo, John T. Reager, James S. Famiglietti, Katalyn Voss, Sean Swenson, and Matthew Rodell. 2015. "Quantifying Renewable Groundwater Stress with GRACE." *Water Resources Research* 51 (7): 5217–38.

Rieland, Randy. 2013. "Learning from Nature How to Deal with Nature." Smithsonian.com, http://www.smithsonianmag.com/innovation/learning-from-nature-how-to-deal-with-nature-5005595/.

Rignot, E., Jeremie Mouginot, Mathieu Morlighem, Helene Seroussi, and B. Scheuchl. 2014. "Widespread, Rapid Grounding Line Retreat for Pine Island, Thwaites, Smith and Kohler Glaciers, West Antarctica, from 1992 to 2011." *Geophysical Research Letters* 41 (10): 3502–09.

Riordan, Michael, and Lillian Hoddeson. 1997. Crystal Fire: The Invention of the Transistor and the Birth of the Information Age (Sloan Technology Series). New York: Norton.

Roberts, Callum. 2012. *Ocean of Life: The Fate of Man and the Sea.* London: Penguin.

Roberts, J. Timmons. 2011. "Multipolarity and the New World (Dis)Order: US Hegemonic Decline and the Fragmentation of the Global Climate Regime." *Global Environmental Change* 21 (3): 776–84.

Rockström, Johan, and Malin Falkenmark. 2015. "Agriculture: Increase Water Harvesting in Africa." *Nature* 519 (7543): 283–85.

Rockström, Johan, Will Steffen, Kevin Noone, Asa Persson, F. Stuart Chapin III, Eric F. Lambin, Timothy M. Lenton, et al. 2009. "A Safe Operating Space for Humanity." *Nature* 461: 472–75.

Rodriguez, Antonio B., F. Joel Fodrie, Justin T. Ridge, Niels L. Lindquist, Ethan J. Theuerkauf, Sara E. Coleman, Jonathan H. Grabowski, et al. 2014. "Oyster Reefs Can Outpace Sea-Level Rise." *Nature Climate Change* 4: 493–97.

Rogers, Alex, and Daniel D'A. Laffoley. 2011. *International Earth System Expert Workshop on Ocean Stresses and Impacts.* IPSO Summary Report. Oxford, UK: International Programme on the State of the Ocean.

Romm, Joseph. 2011. "Desertification: The Next Dustbowl." *Nature* 478 (7370): 450–51.

Rong, Jiang. 2009. *Wolf Totem.* London: Penguin.

Rosenau, James N. 1995. "Governance in the Twenty-First Century." *Global Governance* 1 (1): 13–43.

Rotillon, G. 2010. *Economie des ressources naturelles.* Paris: La Découverte.

Royal Society. 2009. "Reaping the Benefits: Science and the Sustainable Intensification of Global Agriculture." Royal Society Document 11/09. London: Royal Society.

Rupp, David E., Philip W. Mote, Neil Massey, Cameron J. Rye, Richard Jones, and Myles R. Allen. 2012. " Did Human Influence on Climate Make the 2011 Texas Drought More Probable?" *Bulletin of the American Meteorological Society* 93: 1052–67.

Rushdie, Salman. 1981. *Midnight's Children.* New York: Penguin.

Sandford, Rosemary. 1996. "International Environmental Treaty Secretariats: A Case of Neglected Potential?" *Environmental Impact Assessment Review* 16 (1): 3–12.

Sarewitz, Daniel, and Richard Nelson. 2008. "Three Rules for Technological Fixes." *Nature* 456: 871–72.

Savage, Leonard J. 1954. *The Foundations of Statistics.* New York: John Wiley.

Schafer Meredith G., Andrew A. Ross, Jason P. Londo, Connie A. Burdick, E. Henry Lee, Steven E. Travers, Peter K. Van de Water, and Cynthia L. Sagers. 2011. "The Establishment of Genetically Engineered Canola Populations in the U.S." *PLoS ONE* 6 (10): e25736.

Schelling, Thomas C. 1981. *The Strategy of Conflict.* Cambridge, MA: Harvard University Press.

Schiavo, Joseph. 2011. "États-Unis: Le Président qui Prenait le Climat au Sérieux." In *Regards sur la Terre 2011*, ed. Pierre Jacquet, Rajendra Kumar Pachauri, and Laurence Tubiana. Paris: Armand Colin.

Schmidheiny, Stephan. 1992. *Changing Course: A Global Business Perspective on Development and the Environment.* Boston: MIT Press.

Scholte, Jan Aart. 2002. "Civil Society and Democracy in Global Governance." *Global Governance* 8 (3): 281–304.

Schumpeter, Joseph A. 1911. *Theorie der Witschaftlichen Entwicklung*, Wien; translated into English as "The Theory of Economic Development" Cambridge, MA: Harvard University Press.

Selin, Noelle. 2013. "Forty Years of International Mercury Policy: The 2000s and Beyond (Part 3 of 3)" (blog). Mercury Science and Policy at MIT. http://mercury policy.scripts.mit.edu/blog/?p=395.

Selin, Noelle, and Henrik Selin. 2006. "Global Politics of Mercury Pollution: The Need for Multi-Scale Governance." *Review of European, Comparative, and International Environmental Law* 15 (3): 258–69.

Sen, Amartya. 1983. *Poverty and Famines: An Essay on Entitlement and Deprivation.* Oxford, UK: Oxford University Press.

Seong-kyu. 2013. "Audition du Nouveau Ministre de l'Environnement." *Korea Herald.*

Shapiro, Carl. 2000. "Navigating the Patent Thicket: Cross Licenses, Patent Pools, and Standard-Setting." In *Innovation Policy and the Economy* (vol. 1), ed. Adam B. Jaffe, Josh Lerner, and Scott Stern, 119–50. Cambridge, MA: MIT Press.

Smith, Adam. 1776. *An Inquiry Into the Nature and Causes of the Wealth of Nations.* London: Methuen.

Sorg, Annina, Tobias Bolch, Markus Stoffel, Olga Solomina, and Martin Beniston. 2012. "Climate Change Impacts on Glaciers and Runoff in Tien Shan." *Nature Climate Change* 2: 725–31.

Stern, Nicholas. 2006. *The Stern Review on the Economics of Climate Change.* Report commissioned by the Government of the United Kingdom. London: HM Treasury.

——. 2011. "A Profound Contradiction at the Heart of Climate Change Policy." *Financial Times*, December 9. https://www.ft.com/content/52f2709c-20f0-11e1-8a43-00144feabdc0.

——. 2015. "*Why Are We Waiting? The Logic, Urgency, and Promise of Tackling Climate Change.*" Cambridge MA: MIT Press.

Sterner, Thomas, ed. 2011. *Fuel Taxes and the Poor: The Distributional Effects of Gasoline Taxation and Their Implications for Climate Policy.* Washington, DC: RFF Press.

Stewart, Heather, and Larry Elliott. 2013. "Nicholas Stern: 'I Got It Wrong on Climate Change—It's Far, Far Worse.'" *The Observer*, January 26. https://www.theguardian.com/environment/2013/jan/27/nicholas-stern-climate-change-davos.

Stiglitz, Joseph E. 2006. *Making Globalization Work*. New York: W.W. Norton.

Stokstad, Erik. 2012. "Field Research on Bees Raises Concern About Low-Dose Pesticides." *Science* 335 (6076): 1555–56.

Sukhdev, Pavan. 2010. "Putting a Value on Nature Could Set Scene for True Green Economy." *The Guardian*, February 10. https://www.theguardian.com/commentisfree/cif-green/2010/feb/10/pavan-sukhdev-natures-economic-model.

Sumaila, Ussif Rashid. 2013. *Game Theory and Fisheries: Essays on the Tragedy of Free For All Fishing.* London: Routledge.

Sumaila, Ussif Rashid, Vicky Lam, Frédéric le Manach, Wilf Swartz, and Daniel Pauly. 2013. *Global Fisheries Subsidies*. Report created for European Parliament's Committee on Fisheries. Brussels: Directorate-General for Internal Policies.

Surendra, K. C., Devin Takara, Andrew G. Hashimoto, and Samir Kumar Khanal. 2014. "Biogas as a Sustainable Energy Source for Developing Countries: Opportunities and Challenges." *Renewable and Sustainable Energy Reviews* 31: 846–59.

Sutton, Mark A., and Albert Bleeker. 2013. "Environmental Science: The Shape of Nitrogen to Come." *Nature* 494 (7438): 435–37.

ter Steege, Hans, Nigel C. A. Pitman, Timothy J. Killeen, William F. Laurance, Carlos A. Peres, Juan Ernesto Guevara, Rafael P. Salomão, et al. 2015. "Estimating the Global Conservation Status of More Than 15,000 Amazonian Tree Species." *Science Advances* 1 (1): e1500936.

Tilghman, Syreeta L, Melyssa R. Bratton, H. Chris Segar, Elizabeth C. Martin, Lyndsay V. Rhodes, Meng Li, John A. McLachlan, Thomas E. Wiese, Kenneth P. Nephew, and Matthew E. Burrow. 2012. "Endocrine Disruptor Regulation of MicroRNA Expression in Breast Carcinoma Cells." *PLoS ONE* 7 (3): e32754.

Tirole, J. 2003. *Protection de la Propriété Intellectuelle: Une Introduction et Quelques Pistes de Réflexion.* Report for the Conseil d'Analyse Economique, Premier Ministre. Paris: Documentation Française.

Toensmeier, Eric, ed. 2016. *The Carbon Farming Solution: A Global Toolkit of Perennial Crops and Regenerative Agriculture Practices for Climate Change Mitigation and Food Security.* White River Junction, VT: Chelsea Green.

Tollefson, Jeff. 2013. "Methane Leaks Erode Green Credentials of Natural Gas." *Nature* 493 (7430): 12.

Tom, W. K., and J. A. Newberg. 1998. "U.S. Enforcement Approaches to the Antitrust–Intellectual Property Interface." In *Competition Policy and*

Intellectual Property Rights in the Knowledge-Based Economy, ed. Robert D. Anderson and Nancy T. Gallini, 343–93. Calgary, Alberta, Canada: University of Calgary Press.

Torres, Alejandra. 2012. Director of International Affairs, Ministry of Environment and Sustainable Development of Colombia and one of the leading negotiators. (Stockholm Environment Institute, briefs).

Townes, Charles H. 1999. *How the Laser Happened: Adventures of a Scientist.* Oxford, UK: Oxford University Press.

Trucost.com. 2011. "The State of Green Business 2011." https://www.trucost.com/publication/state-green-business-2011/.

Tubiana, Laurence, and Tancrede Voituriez. 2007. "Emerging Powers in Global Governance: New Challenges and Policy Options." Paper prepared for the Emerging Powers in Global Governance: New Challenges and Policy Options Conference, Paris, July 6–7.

Tubiana, Laurence, and Matthieu Wemaëre. 2009. "Climate Change Action: Can the EU Continue to be a Leader?" In *The EU in a World in Transition: Fit for What Purpose?* ed. Loukas Tsoukalis. London: Policy Network.

UN Department of Economic and Social Affairs. 2013. "Voluntary Commitments and Partnerships for Sustainable Development." *Sustainable Development in Action* 1.

UN Development Programme. 2006. *Human Development Report 2006*. New York: Palgrave Macmillan.

UN Development Programme. 2007. *Human Development Report 2007*. New York: Palgrave Macmillan.

UN Environment Programme. 2005. "Action on Heavy Metals Among Key GC Decisions" (press release, February 25). http://staging.unep.org/Documents.Multilingual/Default.asp?DocumentID=424&ArticleID=4740&l=en.

UN General Assembly. 2011. "Accelerating Progress Towards the Millennium Development Goals: Options for Sustained and Inclusive Growth and Issues for Advancing the United Nations Development Agenda Beyond 2015." Sess. 66, A/66/126.

United Nations. 1972. "Declaration of the United Nations Conference on the Human Environment." http://www.un-documents.net/unchedec.htm.

——. 2012. "Rio+20 Conference on Sustainable Development: The Future We Want." https://rio20.un.org/sites/rio20.un.org/files/a-conf.216l-1_english.pdf.pdf.

U.S.-China Climate Change Working Group. 2014. "Report of the U.S.-China Climate Change Working Group to the 6th Round of the Strategic and Economic Dialogue July 9, 2014."

U.S. Department of Defense. 2014. *2014 Climate Change Adaptation Roadmap*. Washington, DC: Government Printing Office.

U.S. Department of Energy. 2009. "States with Renewable Portfolio Standards."

U.S. Federal Trade Commission. 2003. *To Promote Innovation: The Proper Balance of Competition and Patent Law and Policy*. Washington, DC: Federal Trade Commission.

U.S. Global Change Research Program. 2014. National Climate Assessment 2014. http://nca2014.globalchange.gov/.

van Noorden, Richard. 2014. "The Rechargeable Revolution: A Better Battery." *Nature* 507: 26–28.

Vasconcelos, Vitor V., Francisco C. Santos, and Jorge M. Pacheco. 2013. "A Bottom-Up Institutional Approach to Cooperative Governance of Risky Commons." *Nature Climate Change* 3: 797–801.

Venter, J. Craig. 2013. *Life at the Speed of Light: From the Double Helix to the Dawn of Digital Life*. New York: Viking.

Victor, David G. 2001. *The Collapse of the Kyoto Protocol and the Struggle to Slow Global Warming*. Princeton, NJ: Princeton University Press.

Vogler, John. 2002. "In the Absence of the Hegemon: EU Actorness and the Global Climate Change Regime." National Europe Center Paper No. 20, presented at the Conference on the European Union and International Affairs, National Europe Centre, Australian National University, July 3–4, 2002.

Vogler, John, and Charlotte Bretherton. 2006. "The European Union as a Protagonist to the United States on Climate Change." *International Studies Perspectives* 7 (1): 18.

Vogler, John, and Hannes Stephan. 2007. "The European Union in Global Environmental Governance: Leadership in the Making?" *International Environmental Agreements: Politics, Law and Economics* 7: 389–413.

von Frantzius, Ina. 2004. "World Summit on Sustainable Development Johannesburg 2002: A Critical Analysis and Assessment of the Outcomes." *Environmental Politics* 13 (2): 467–73.

von Hippel, Eric. 2005. *Democratizing Innovation*. Cambridge, MA: MIT Press.

von Neumann, John, and Oskar Morgenstern. 1944. *Theory of Games and Economic Behavior*. Princeton, NJ: Princeton University Press.

Wallström, Margaret. 2001. "The Results of the Göteborg European Council in Respect to Sustainable Development and Climate Change." Brussels: Centre for European Policy Studies, June 18.

Wang, Wenkin, G. Haberer, H. Gundlach, C. Gläber, T. Nussbaumer, M. C. Luo, A. Lomsadze, et al. 2014. "The *Spirodela polyrhiza* Genome Reveals Insights into Its Neotenous Reduction Fast Growth and Aquatic Lifestyle." *Nature Communications* 5.

Ward, Barbara, and René Dubos. 1983. *Only One Earth: The Care and Maintenance of a Small Planet*. New York: Norton.

Watson, David M., and Matthew Herring. 2012. "Mistletoe as a Keystone Resource." *Proceedings of the Royal Society* (Series B) 279: 3853–60.

Watson, James D. and Francis Crick. 1953. "A Structure for Deoxyribose Nucleic Acid", *Nature* 171: 737–38.

Watt, Robert. 2012. "Do We Need a New Set of Development Goals?" *SEI Bulletin*, November 21. http://www.sei-international.org/-news-archive/2510-do-we-need-a-new-set-of-global-development-goals.

Watts, Nick, W. Neil Adger, Paolo Agnolucci, Jason Blackstock, Peter Byass, Wenijia Cai, Sarah Chaytor, et al. 2015. "Health and Climate Change: Policy Responses to Protect Public Health." *The Lancet Commissions* 386 (10006): 1861–1914.

Weber, Steven. 2005. *The Success of Open Source.* Cambridge, MA: Harvard University Press.

White House Office of the Press Secretary. 2015. "U.S.-China Joint Presidential Statement on Climate Change." https://obamawhitehouse.archives.gov/the-press-office/2015/09/25/us-china-joint-presidential-statement-climate-change.

Wilcox, Chris, Erik van Sebille, and Britta Denise Hardesty. 2015. "Threat of Plastic Pollution to Seabirds Is Global, Pervasive, and Increasing." *Proceedings of the National Academy of Science* 112 (38): 11899–904.

Winkelmann, Ricarda, Anders Levermann, Andy Ridgwell, and Ken Caldeira. 2015. "Combustion of Available Fossil Fuel Resources Sufficient to Eliminate the Antarctic Ice Sheet." *Science Advances* 1 (8): e1500589.

Witting, Jan. 2016. "The Phoenix Islands Protected Area: Lessons for the Stewardship of the Global Ocean." Lecture at Sea Education Association, Woods Hole, MA, March 20.

Woodcock, Ben A., Nicholas J. B. Isaac, James M. Bullock, David B. Roy, David G. Garthwaite, Andrew Crowe, and Richard F. Pywell. 2016. "Impact of Neonicotinoid Use on Long-Term Population Changes in Wild Bees in England." *Nature Communications* 7.

Wootton, David. 2015. *The Invention of Science: A New History of the Scientific Revolution.* London: Allen Lane.

World Bank. 2008. *World Development Report 2008: Agriculture for Development.* Washington, DC: World Bank.

World Bank. 2012. "Turn Down the Heat—Why a 4°C Warmer World Must Be Avoided. A Report for the World Bank by the Potsdam Institute for Climate Impact Research and Climate Analysis." Washington, DC: World Bank.

——. 2017. *The Sunken Billions Revisited: Progress and Challenges in Global Marine Fisheries* (Environment and Sustainable Development series). Washington, DC: World Bank Publications.

Worldwatch Institute. 2015. *Growing Land Grabs.* Washington, DC: Island Press.

World Wildlife Fund. 2016. *WWF Living Planet Report 2016: Risk and Resilience in a New Era.* Gland, Switzerland: WWF International.

Wouters, Bert, Alba Martín-Español, Velt Helm, Thomas Flament, J. M. van Wessem, Stefan R. M. Ligtenberg, Michiel Roland Van den Broeke, and Jonathan L. Bamber. 2015. "Dynamic Thinning of Glaciers on the Southern Antarctic Peninsula." *Science* 348 (6237): 899–903.

Wu, Yuxuan, Dan Liang, Yinghua Wang, Meizhu Bai, Wei Tang, Shiming Bao, Zhiqiang Yan, Dangsheng Li, and Jinsong Li. 2013. "Correction of a Genetic Disease in Mouse via Use of CRISPR-Cas9." *Cell Stem Cell* 13 (6): 659–62.

Xiaolong, Qiu. 2012. *Don't Cry, Tai Lake.* New York: St. Martin's.

Yang, Chi-Jen, and Robert B. Jackson. 2013. "China's Synthetic Natural Gas Revolution." *Nature Climate Change* 3: 852–54.

Yao, Tandong, Lonnie Thompson, Wei Yang, Wusheng Yu, Yang Gao, Xuejun Guo, Xiaoxin Yang, et al. 2012. "Different Glacier Status with Atmospheric Circulations in Tibetan Plateau and Surroundings." *Nature Climate Change* 2: 663–67.

Yu, Junsheng, Yifan Zheng, and Jiang Huang. 2014. "Towards High Performance Organic Photovoltaic Cells: A Review of Recent Development in Organic Photovoltaics." *Polymers* 6 (9): 2473–509.

Zhang, Wei-Feng, Zheng-xia Dou, Pan He, Xiao-Tang Ju, David Powlson, Dave Chadwick, David Norse, et al. 2013. "New Technologies Reduce Greenhouse Gas Emissions from Nitrogenous Fertilizer in China." *Proceedings of the National Academy of Sciences* 110 (21): 8375–80.

Zhang, Yuning, Ningning Tang, Yuguang Niu, and Xiaoze Du. 2016. "Wind Energy Rejection in China: Current Status, Reasons and Perspectives." *Renewable and Sustainable Energy Reviews* 66: 322–44.

Zhu, Zhi, Akihiro Kushima, Zongyou Yin, Lu Qi, Khalil Amine, Jun Lu, and Ju Li. 2016. "Anion-Redox Nanolithia Cathodes for Li-Ion Batteries." *Nature Energy* 1. doi: 10.1038/nenergy.2016.111.

Ziman, John. 2000. *Real Science: What It Is, and What It Means.* Cambridge, UK: Cambridge University Press.

INDEX